Vikas Dubey, Sri R. K. Mishra, Marta Michalska-Domańska, Vaibhav De
Water Resource Technology

Also of Interest

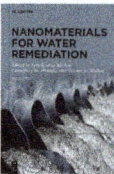

Nanomaterials for Water Remediation
Ajay Kumar Mishra, Chaudhery M. Hussain, Shivani B. Mishra (Eds.),
2020
ISBN 978-3-11-064336-7, e-ISBN 978-3-11-065060-0

Drinking Water Treatment.
An Introduction
Eckhard Worch, 2019
ISBN 978-3-11-055154-9, e-ISBN 978-3-11-055155-6

Aquatic Chemistry.
For Water and Wastewater Treatment Applications
Ori Lahav, Liat Birnhack, 2019
ISBN 978-3-11-060392-7, e-ISBN 978-3-11-060395-8

Environmental Engineering.
Basic Principles
Vesna Tomašić, Bruno Zelić (Eds.), 2018
ISBN 978-3-11-046801-4, e-ISBN 978-3-11-046803-8

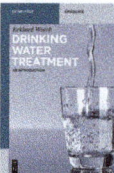

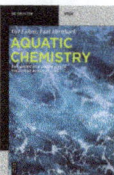

Water Resource Technology

Technology

—

Management for Engineering Applications

Edited by
Vikas Dubey, Sri R. K. Mishra, Marta Michalska-Domańska,
Vaibhav Deshpande

DE GRUYTER

Editors

Dr. Vikas Dubey
Bhilai Institute of Technology
Abhanpur Road, Atal Nagar
Raipur, Chhattisgarh 493661
India
jsvikasdubey.physics@gmail.com

Dr. R. K. Mishra
Central Avenue Road
2 Cross Street
Ananda Nagar 490020
India
Rkmishra4776@gmail.com

Dr. Marta Michalska-Domańska
Military University of Technology
Military Institute of Optoelectronics
Urbanowicza Str 2
00-908 Warsaw
Poland
marta.michalska@wat.edu.pl

Asst. Prof. Vaibhav Deshpande
Bhilai Institute of Technology
Department of Civil Engineering
Raipur, Chhattisgarh
India
dvaibhav86@bitraipur.ac.in

ISBN 978-3-11-072134-8
e-ISBN (PDF) 978-3-11-072135-5
e-ISBN (EPUB) 978-3-11-072155-3

Library of Congress Control Number: 2021935411

Bibliographic information published by the Deutsche Nationalbibliothek
The Deutsche Nationalbibliothek lists this publication in the Deutsche Nationalbibliografie;
detailed bibliographic data are available on the Internet at http://dnb.dnb.de.

© 2021 Walter de Gruyter GmbH, Berlin/Boston
Cover image: Mara Duchetti/iStock/Getty Images Plus
Typesetting: Integra Software Services Pvt. Ltd.
Printing and binding: CPI books GmbH, Leck

www.degruyter.com

Contents

Ashay Shende, Mahendra Umare
1 Analysis of rainfall and runoff of a catchment area using HYMOS —— 1

Ayush Ransingh, Anjali Khambete, Sumantra Chaudhuri
2 An investigation on hydrodynamic cavitation as pre-treatment
  or post-treatment technology for treating CETP wastewater —— 9

Rubina Sahin, Anil Kumar, Rituraj Chandrakar,
Marta Michalska-Domańska, Vikas Dubey
3 The physico-chemical interaction of fluorine with the environment —— 17

Ghanshyam Shakar, Bhumika Das
4 The effect of chemical waste produced by industrial area near Raipur
  (Chhattisgarh) on the quality of drinking water —— 39

Pallavi Pradeep Khobragade, Ajay Vikram Ahirwar
5 Seasonal variation of PM_{10} in the ambient air over an urban industrial
  area —— 55

Sahajpreet Kaur Garewal, Avinash D. Vasudeo
6 Groundwater sustainability assessment using multi-criteria analysis —— 67

Ankit Shinde, Vaibhav P. Deshpande
7 Impact of sodium absorption ratio on Kharun river basin, Raipur,
  Chhattisgarh —— 77

Mahendra Umare, Ashay D. Shende, Valsson Varghese, A. M. Badar
8 Experimental approach to study the fall velocity of different sized particle
  in quiescent liquid column —— 85

Shubham Chandrakar, Manish Kumar Sinha
9 Comparative analysis of NDVI and LST to identify Urban Heat Island effect
  using remote sensing and GIS —— 99

Index —— 111

Ashay Shende, Mahendra Umare
1 Analysis of rainfall and runoff of a catchment area using HYMOS

Abstract: The primary source of water in India is rainfall which is experienced over most part of the country. The rainfall varies from place to place and is measured using more than 3,000 rain gauge stations, which are set up by Indian Meteorological Department (IMD) and Water Resource Department of State Government. The average annual rainfall of India is about 119.4 cm. According to Irrigation Commission of India, the total annual average precipitation in India is about 4,000 km^3, which contributes towards a total annual surface water flow of about 1,869 km^3 and the remaining is lost as evaporation, transpiration and infiltration. If the intensity of rainfall is more, it can cause hazardous flood leading to high casualties and property damages. To mitigate the floods, structural and non-structural measures are inevitable along with strong flood forecasting or early warning systems. This is possible when effective and scientific analysis of rainfall and runoff is carried out for a vulnerable catchment area. This chapter emphasizes the application of rainfall–runoff relationship and Unit Hydrograph in Flood Forecasting using a software tool HYMOS. The results obtained show good amount of accuracy and corroborate with site parameters. For calculation purpose, the data of catchment area serving Multipurpose (Irrigation and Hydroelectric) Project Pench in Maharashtra–Madhya Pradesh is used, and rainfall is calculated for full climatic station, Kamthikhairy (latitude 21°22′01″N and 79°11′37″E) with altitude 295.00 m using various climatologically factors and HYMOS tool. This will provide data for the planning, scheduling of different irrigation and non-irrigational projects, flood routing and so on of the region.

1.1 Introduction

Long-term planning of water resource project depends upon the correlation of rainfall and surface hydrology of respective catchment areas. The analysis helps the concerned department to understand the availability of water for the use in various sectors (Kebila, 2008).

Acknowledgement: We are grateful to the Hydrology Division, Water Resource Department, Nagpur, Government of Maharashtra, India, for providing data and necessary facilitation of analysis on HYMOS.

Ashay Shende, Mahendra Umare, Department of Civil Engineering, KDK College of Engineering, Nagpur 440009, Maharashtra, India, e-mail: ashayshende220@gmail.com

https://doi.org/10.1515/9783110721355-001

Precipitation is measured using various pluviometers (types of rain gauges). The characteristic features need to be considered are catchment area and topography, intensity and type of storm, and season. For studying the characteristic features of rainfall and runoff, all the meteorological parameters should be considered. All the components of hydrological cycle (such as runoff, surface runoff, sub-surface runoff, precipitation, evaporation, condensation, transpiration, infiltration) need to be evaluated for analysing discharge of particular catchment (Kim and Kim, 2020). The meteorological parameters are recorded using different instruments, which provide information on local weather and climatic conditions.

All the recorded qualitative and quantitative data pertaining to meteorological and hydrological parameters are fed into Surface Water Data Entry System (SWDES) software. HYMOS is used to analyse the data fed to SWDES thereby helping the officials to achieve effective reservoir operation.

Hence, the user-friendly features of HYMOS make it a versatile tool for carrying out research and consultancy in the field of water management. At present, HYMOS 4.03 version is in use. Using statistical data from the water resource department, various analyses such as Unit Hydrograph Ordinate (UHO), flood forecasting, coefficient of rainfall and run-off and its relationship, identification of characteristic features of catchment, surface water analysis are performed.

1.1.1 Background

Many researchers have studied rainfall–runoff relation for different catchment, watershed for effective computation of flood discharge, planning and management of water resource projects and so on. However, a few notable research have been reviewed and summarized in Table 1.1 as given below:

Table 1.1: Review of literature.

SN	Name of paper	Review
01	Kebila (2008)	This article focused on developing a rainfall–runoff model using limited data and characteristics of catchment.
02	Verma (2005)	The authors have carried out work on rainfall, runoff and sediment losses from watersheds of the central Himalayas over different land cover. Precipitation was recorded for five rain gauge stations spread into different regions.
03	Nur-Sharifah (2015)	The research was conducted to analyse the correlation between the precipitation and surface flow was modelled for the Lipis River using Hydrological Modeling System (HEC-HMS). The data collected were used to prevent flash flood and insufficient capacity of drainage problem.

Table 1.1 (continued)

SN	Name of paper	Review
04	Merz and Blöschl (2009)	The research was carried out to understand the spatio-temporal variation in the runoff coefficient and was observed that the antecedent soil moisture and extent of catchment are governing factors for the same.
05	Linsley (1967)	Data pertaining to seasonal and yearly rainfall over the years were analysed to obtain a correlation between rainfall and runoff. The authors concluded that the runoff is about one sixth of the rainfall.
06	Straub and Melching (2000)	The authors have developed an equation for estimating the time of concentration and storage coefficient of the Clark unit hydrograph for a small watershed.
07	Ahmad et al. (2009)	The authors have opined that hydrologic response of the catchment is closely related to the ratio which is around 0.5 showing runoff diffusion and translation flow effects that play equivalent role in slow hydrologic response.
08	Loague and Freeze (1985)	The authors have explored the predictive power of runoff from the rainfall using different mathematical models, namely Regression, Unit Hydrograph and Quasi-physical-based model over a small upland catchment. They concluded that all the three models are having low predictive power, but the regression model is better than other two models.
09	Jaafar et al. (2016)	The authors have revisited the work done by earlier researchers in establishing the rainfall–runoff correlation and flood modelling. They opined that the knowledge of correlation between rainfall–runoff is critical in forecasting and managing the flood event.
10	Ahmad et al. (2009)	GIS technique was used to get the instantaneous unit hydrograph (IUH) wherein the basin characteristics were obtained using GIS. Comparison of the analytical and observed values was done and found to be matching well.
11	Kim and Kim (2020)	The researchers analysed the effect of distributed precipitation event on rainfall and runoff thereafter over a watershed area. However, many other factors too affect the rainfall–runoff relation.
12	Sabol (1988)	R-value was evaluated using the hydrograph recession technique on the recorded hydrographs and an empirical equation was suggested to compute the R-value for an un-gauged watershed.

1.1.2 Study area

The basin under the study was Kamthikhairi catchment (latitude 21°22′01″ N and 79°11′37″ E, altitude 295.00 m above MSL) covering 4,825.0 km^2 area and served by the Pench River, a tributary of the Kanhan River in the Godavari Basin. The Pench River boasts two major projects, namely Irrigation and Hydroelectric.

1.2 Methodology

1.2.1 Data collection

The meteorological data were collected from water resources and hydrology department of Nashik from 2011 to 2015, which comprises the following parameters such as:
1. Sunshine hours
2. Relative humidity or dry bulb and wet bulb temperature
3. Rainfall
4. Air temperature
5. Wind run

1.2.2 Software used

a) SWDES
b) HYMOS

a) SWDES (Surface Water Data Entry System)

SWDES has been developed by Dwelft Hydraulics, the Netherland, to suit Indian conditions. Use of the software tool is for data entry and primary validation at sub-divisional level. Validation is a process to ensure that readings entered in software resemble those taken in the field.

b) HYMOS (Hydrological Modelling Software)

HYMOS is a time-series oriented soft-computing tool to process, analyse and validate large hydro-environmental and meteorological data to obtain hydrographs and relation between rainfall and runoff.

1.2.3 Data analysis

The data from the Kamthikhairi full climatic station were analysed to obtain a unit hydrograph, which is then used to get the direct runoff hydrograph. Following are the unit hydrographs obtained for the area under study over the years using the soft-computing tool.

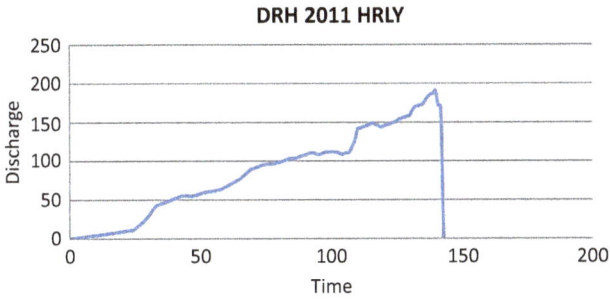

Figure 1.1: Unit hydrograph 2011.

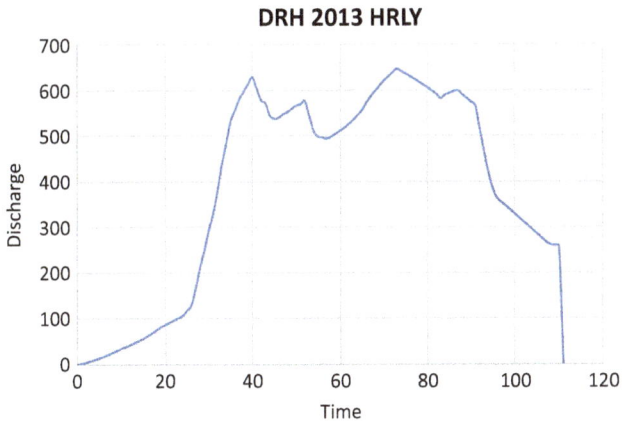

Figure 1.2: Unit hydrograph 2013.

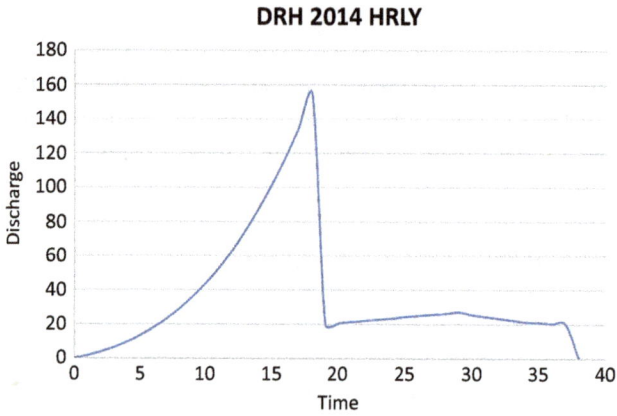

Figure 1.3: Unit hydrograph 2014.

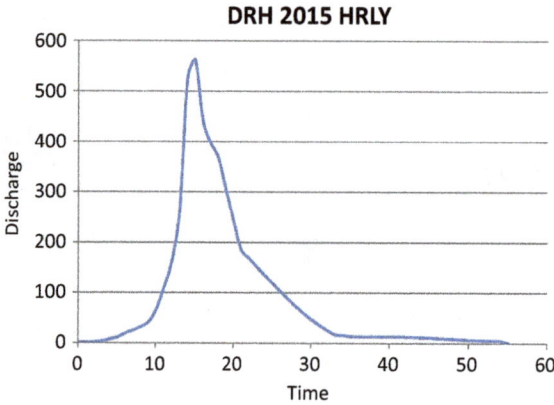

Figure 1.4: Unit hydrograph 2015.

Annual rainfall depth (mm).

SN	Year	Annual rainfall depth (mm)
1	2011	1,140.42
2	2013	3,869.10
3	2014	2,167.01
4	2015	3,293.00

Monthly percentage of HRD to MPC.

SN	Month	Meteorological precipitation computation (MPC)	Hydrological runoff depth (HRD)	HRD in %
1	JUN	416.4	8.217205	1.973392
2	JUL	739.3	160.7519	21.7438
3	AUG	420.4	337.0683	80.178
4	SEP	88.6	33.08576	37.34284
5	OCT	116.6	61.81306	53.01292

1.3 Conclusion

The hydrological runoff depth (HRD) found to be more in the month of August for meteorological precipitation computation (MPC) of corresponding month, which indicates the losses due to infiltration are less. Though the MPC for the month of July comes out to be more but due to the losses, the HRD is obtained less as comparative to other months of monsoon period. The same is depicted for the month of June, which indicates higher losses. Based on the values of HRD for various months of monsoon period, reservoir operation and planning of supply of water for various needs such as irrigation, drinking and so on can be hydro-meteorological data to obtained unit hydrograph, which will help in further water resource management.

References

Ahmad, M.M., Ghumman, A.R., Ahmad, S. (2009). Estimation of Clark's instantaneous unit hydrograph parameters and development of direct surface runoff hydrograph. Water Resources Management, 23(12), 2417–2435.

Jaafar, K., et al. 2016. "A review on flood modelling and rainfall-runoff relationships." Proceedings – 2015 6th IEEE Control and System Graduate Research Colloquium, ICSGRC 2015: 158–162.

Kebila, B.S. (2008). Analysis of the rainfall-runoff pattern of a catchment with limited data to estimate the runoff potential. Department of Technical Water Resources (TVVR 08/5001-73): 1–59.

Kim, C., Kim, D.H. (2020). Effects of rainfall spatial distribution on the relationship between rainfall spatiotemporal resolution and runoff prediction accuracy. Water (Switzerland), 12, 3.

Linsley, R.K. (1967). The relation between rainfall and runoff: Review paper. Journal of Hydrology, 5, 297–311. http://dx.doi.org/10.1016/S0022-1694(67)80128-8.

Loague, K.M., Freeze, R.A. (1985). A comparison of rainfall-runoff modeling techniques on small upland catchments. Water Resources Research, 21(2), 229–248.

Merz, R., Blöschl, G. (2009). A regional analysis of event runoff coefficients with respect to climate and catchment characteristics in Austria. Water Resources Research, 45(1), 1–19.

Nur Sharifah Ayu Binti Salim, (2015), Study Of Rainfall-Runoff Relationship Using Hydrologic Modeling System (Hec-Hms) For Lipis River, UG Thesis University Malaysia Pahang.

Sabol, G. V. (1988). Clark Unit Hydrograph and R-Parameter Estimation. Journal of Hydraulic Engineering, 114(1), 103–111.

Straub, Timothy D., Melihing, CS., Kocher, Kyle E. (2000). Equation for estimating clark unit hydrograph parameters for small rural watersheds in illinois. U.S. Dept. of the Interior, U.S. Geological Survey.

Verma, P. K., Kothyari, B. P. (2005). Assessment of rainfall, runoff and sediment losses of the Bheta Gad watershed, India, Resource constraints and management options in mountain watersheds of the Himalayas. Proceedings of a Regional Workshop, Kathmandu, Nepal, pp.155–166.

Ayush Ransingh, Anjali Khambete, Sumantra Chaudhuri

2 An investigation on hydrodynamic cavitation as pre-treatment or post-treatment technology for treating CETP wastewater

Abstract: In this work, hydrodynamic cavitation has been investigated as a pre-treatment and post-treatment technology for in-practice biological treatment technology of Common Effluent Treatment Plant (CETP) wastewater. Requirement of such an investigation arose because the in-practice treatment technology was not able to meet discharge norms. The already established optimal operation parameters of inlet pressure 5 bar and duration of 60 min were used in the operation of hydrodynamic cavitation, whereas the parameters such as MLSS, HRT in biological treatment were kept same as of the actual treatment in plant. The observed parameters are chemical oxygen demand (COD), ammoniacal nitrogen, colour and TDS. In successful 20 trials of hydrodynamic cavitation as a pre-treatment technology, COD, ammonical nitrogen, colour and total dissolved solids (TDS) were reduced to 330 mg/L, 70.8 mg/L, 1,945 hazens and 5.24 ppb, respectively, whereas in 20 successful trials of hydrodynamic cavitation as post-treatment technology, COD, ammonical nitrogen, colour and total dissolved solids (TDS) were reduced to 319 mg/L, 41.44 mg/L, 4,325 hazens and 6.72 ppb, respectively.

2.1 Introduction

With the arrival of new products every other day in the market, the level of luxury is also increasing. Convenience to human life is increasing at the cost of new finished products produced as a result of complex procedures which further lead to even more complex wastewater. Treatment of such wastewater has always remained a tough task with new treatment technologies being tried every now and then. One physical method for treatment of such wastewater is hydrodynamic cavitation.

At experimental level, it was found that hydrodynamic cavitation would yield higher efficiency and would have the capacity to perform better on large scale as compared to ultrasonic cavitation, which was thus seen as a possible alternative (Arrojo and Benito, 2008). Hydrodynaimc cavitation is much more efficient and its scale-up

Ayush Ransingh, Anjali Khambete, Sumantra Chaudhuri, Civil Engineering Department, Sardar Vallabhbhai National Institute of Technology, Surat 395007, Gujarat, India,
e-mail: ransingh.ayush@gmail.com

https://doi.org/10.1515/9783110721355-002

is comparatively easier than the other cavitaion pheonomenon such as acoustics or sonochemical cavitation. At the same time, hydrodynamic cavitation also offers 20–25 times higher rate of degradation and oxidation efficiency (Chakinala et al., 2008).

Real wastewater sample treatment using hydrodynamic cavitation is seldomly worked on and even a very few work have been carried out for the treatment of synthetic sample using hydrodynamic cavitation (Chakinala et al., 2008).

Hydrodynamic cavitation occurs when a liquid is passed through a constriction device such as venture, orifice plate or a throttling valve. During the passage of liquid via throttling device, the velocity increases because the area of flow decreases due to constriction and simultaneously the pressure also decreases. If this decrease in pressure reaches the critical pressure, that is the vapour pressure of liquid, then a large number of cavities are formed. On the downstream side of constriction device, the velocity of flow decreases as the area of flow increases and simultaneously the pressure also increases, and hence the cavities collapse (Gogate and Pandit, 2005).

During treatment of wastewater using hydrodynamic cavitation, two mechanisms occur: first is the thermal decomposition and second is the reaction with free radicals being formed. In thermal decomposition, the volatile component of wastewater gets decomposed due to high temperature region being developed. In reaction with hydroxide radicals, which are strong oxidizing agents, the non-volatile component of wastewater gets oxidized (Saharan et al., 2011).

For a liquid flowing through a pipe, Bernoulli's law holds good and due to constriction device such as orifice plate, the velocity increases because of decreased flow area. Since Bernoulli's law holds good due to increase in velocity, the static pressure of liquid decreases, and hence the cavitation occurs only when the vapour pressure of liquid is achieved, forming a large number of gaseous cavities (Joshi and Gogate, 2012).

Highly toxic and refractory components of wastewater possess a problem for conventional treatment plants worldwide. Since it is difficult to treat these harmful components conventionally, on which most of the treatment plants are based, these are ultimately released into the environment and possess a risk to human health and life, animals and plants. Therfore, it is important to treat these components using different treatment technologies (Mishra and Gogate, 2010). The striking characteristics of cavitation, that is production of free radicals, extreme pressure and temperature conditions can judiciously be used in the interest of wastewater treatment. Sometimes from economy point of view, it may not be acceptable; but for effluents with high COD, hydrodynamic cavitation is strongly recommended (Chakinala et al., 2009).

2.2 Materials and methodology

2.2.1 Materials

Inlet and outlet wastewater in CETP was collected (for secrecy, name of CETP cannot be revealed). The wastewater in CETP is of complex nature being contributed by a number of different industries of varied nature. Since the CETP was facing problem on colour and refractory chemical oxygen demand (COD), these parameters were observed; and since ammonical nitrogen interferes with the refractory COD, ammonical nitrogen was also observed. Eventually, this work gained importance as investigation is based on real wastewater. Since most of the work carried out on hydrodynamic cavitation, which was on synthetic sample, it may or may not indicate the actual performance of hydrodynamic cavitation in real sample.

2.2.2 Experimental setup

The experimental setup used for hydrodynamic cavitation basically consists of a feed tank of 20 L capacity, a multistage centrifugal pump of power rating 1.1 kW and 1.5 HP, control valves, a main line and a bypass line. The main line has flanges which houses the orifice plate of diameter 6 cm with a single hole of 2 mm diameter and 1 mm thickness, used as a cavitating device and the bypass line is used to control the flow through main line. The main line and bypass line should terminate deep inside the tank, atleast below the liquid level so that no air comes in contact with the sample being circulated. The temperature in the feed tank was maintained constant using a cooling bath around the feed tank which results in a variation of ±2° C temperature. The schematic representation of setup is shown in Figure 2.1.

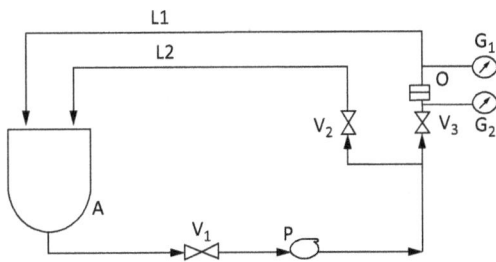

Figure 2.1: Schematic representation of setup.

A = Feed tank   G1 = Inlet Pressure Gauge
L1 = Main line   G2 = Outlet Pressure Gauge
O = Orifice plate   L2 = Bypass line
V1, V2, V3 = Control valve   P = Multistage pump

2.2.3 Methodology

The degradation using hydrodynamic cavitation (HC) will be carried out using circular orifice plate as a cavitating device. Optimization of pressure and treatment time was already performed in previous work (Ransingh and Khambete, 2017). A total of 20 trials were performed for both inlet and outlet wastewater separately, and in each trial 10 L sample was treated under optimized conditions.

In case of inlet wastewater after hydrodynamic cavitation treatment, it is subjected to biological treatment in which aeration time and mixed liquor suspended solids (MLSS) and mixed liquor volatile suspended solids (MLVSS) concentrations were maintained almost similar to what is maintained in CETP because we intend to investigate the suitability of hydrodynamic cavitation as a post or pre-treatment for CETP wastewater and hence we do not want to influence the actual conditions being practised in biological treatment in CETP. As such aeration time of 24 h, MLSS concentration of 4,200–4,800 mg/L and MLVSS concentration of 3,200–3,600 mg/L were maintained. Since Returned Activated Sludge (RAS) of the biological treatment in the plant itself was obtained, no acclimatization period was required. In case of the treatment of inlet wastewater already treated with hydrodynamic cavitation, samples were drawn every 8 h and were analysed for COD and ammonical nitrogen.

In case of outlet wastewater, a total of 20 trials were performed using hydrodynamic cavitation for the samples which were already treated by biological treatment in the treatment plant. Samples were drawn at every 30 min and analysed for ammonical nitrogen and COD.

2.3 Results and discussion

2.3.1 COD reduction in both the treatment

In case of inlet wastewater, initial COD was between 1,100 and 1,700 mg/L, which was reduced to 600–800 mg/L in 60 min of hydrodynamic cavitation treatment. It was further treated in biological treatment where COD was reduced to 400–650 mg/L, 300–500 mg/L and 200–350 in 8 h, 16 h and 24 h of aeration, respectively (graph shown below).

In case of outlet wastewater, initial COD was in the range of 375–575 mg/L, which was reduced to 300–425 mg/L and 250–400 mg/L in 30 and 60 min of hydrodynamic cavitation treatment (graph shown below). Therefore, in previous treatment, 79.40–81.81% reduction was achieved whereas in later case it was 30.43–33.34%.

Chakinala et al. (2009) have shown reduction in total organic carbon (TOC) by around 60% using a combination of hyrodynamic cavitation and Fenton's reagent. Chakinala et al. (2008) have shown reduction in TOC by around 80% using a

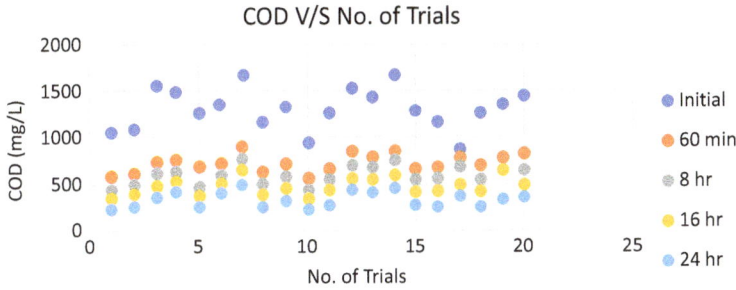

Figure 2.2: COD reduction in HC followed by biological treatment of inlet wastewater.

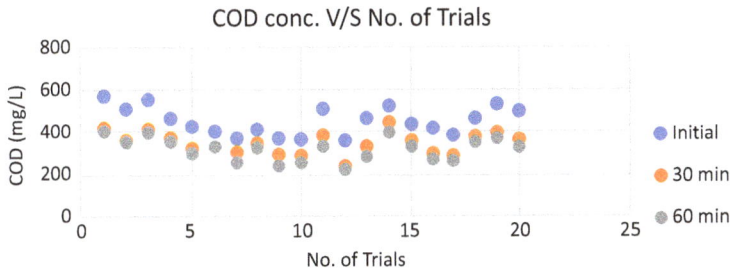

Figure 2.3: COD reduction in HC treatment of outlet wastewater.

combination of hydrodynamic cavitation and advanced Fenton's reagent. Badve et al. (2013) in their work have shown that reduction in COD increases by around 46% on optimal loading of hydrogen peroxide (H_2O_2).

2.3.2 Ammonical nitrogen reduction in both the treatment

In inlet wastewater, the initial ammonical nitrogen concentration was in the range of 140–90 mg/L, which was reduced to 125–70 mg/L, and was further treated with biological treatment where ammonical nitrogen was reduced to 110–65 mg/L, 100–50 mg/L and 90–45 mg/L in 8 h, 16 h and 24 h of aeration, respectively (graph shown below).

In outlet wastewater, the intial ammonical nitrogen concentration was in the range of 85–45 mg/L which was reduced to 60–40, 50–35 mg/L in 30 and 60 min of hydrodynamic cavitation, respectively (graph shown below). Therefore in previous treatment, 35.357–50% reduction was achieved whereas in later case, it was 22.23–41.17%.

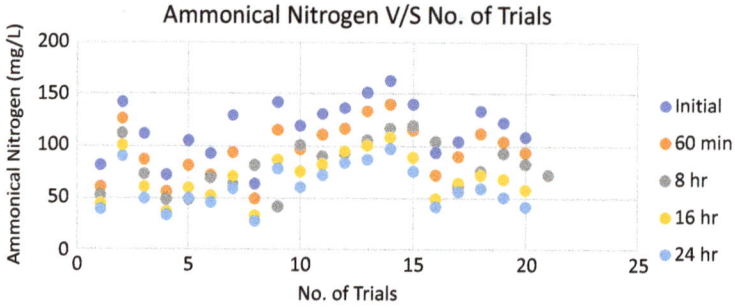

Figure 2.4: Ammonical nitrogen reduction in HC followed by biological treatment of inlet wastewater.

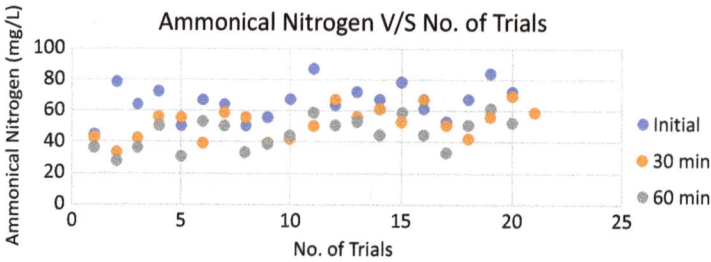

Figure 2.5: Ammonical nitrogen reduction in HC treatment of outlet wastewater.

2.3.3 Colour reduction in both the treatment

In case of inlet wastewater, initial colour was in the range of 3,400–6,000 hazens, which was reduced to 3,000–5,500 hazens in 60 min of hydrodynamic cavitation treatment, and was further reduced in biological treatment to 2,800–4,800 hazens, 2,000–4,000 hazens and 1,800–3,300 hazens in 8 h, 16 h and 24 h aeration time that is approximately 52–55% (graph shown below).

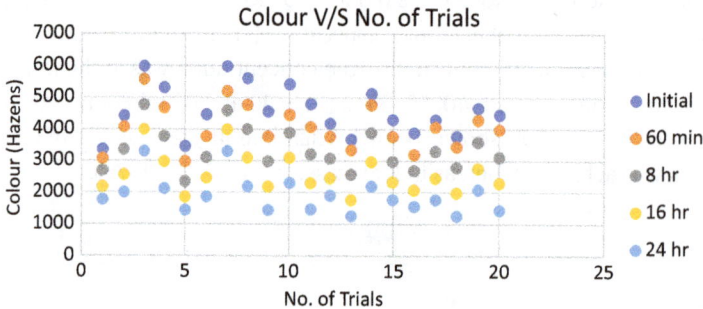

Figure 2.6: Colour reduction in HC followed by biological treatment of inlet wastewater.

In case of outlet wastewater, the intial colour concentration was in the range of 4,000–7,900 hazens which was reduced to 3,200–5,800 hazens and 2,800–5,500 hazens in 30 and 60 min of hydrodynamic cavitation treatment, respectively, which is approximately 30–40% (graph shown below). Therefore in previous treatment, 40–47.09% reduction was achieved whereas in later case, it was approximately 30%.

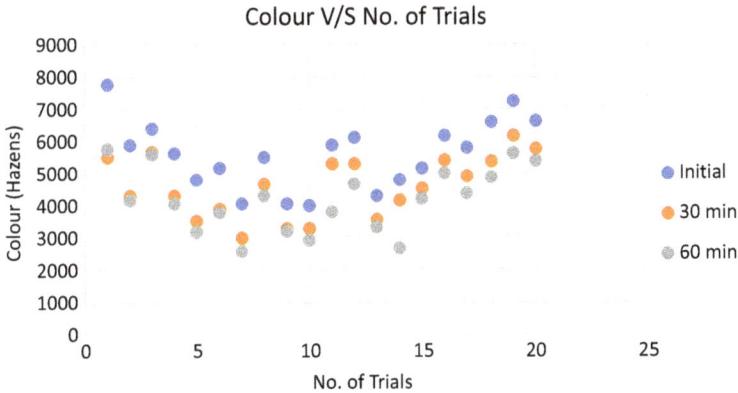

Figure 2.7: Colour reduction in HC treatment of outlet wastewater.

Joshi and Gogate (2012) have shown 91.5% reduction in colour using combination of hydrodynamic cavitation and Fenton's reagent. Gore et al. (2014) have also shown 71.25% reduction in colour using a combination of hydrodynamic cavitation and ozone. Saharan et al. (2011) have also shown significant reduction in colour using a combination of hydrodynamic cavitation and H_2O_2.

2.4 Conclusion

In present work, it was revealed that hydrodynamic cavitation along with biological treatment can effectively degrade Common Effluent Treatment Plant (CETP) wastewater under optimized conditions of operating parameters which are inlet pressure and operation time. The extent of degradation was more when hydrodynamic cavitation was followed by biological treatment. The extent of reduction in COD and ammonical nitrogen was almost similar but reduction in colour was significant. The following points were established during investigation:

i) COD was reduced to 200–350 mg/L in hydrodynamic cavitation followed by biological treatment resulting in 79.40–81.81% reduction.

ii) Ammonical nitrogen was reduced to 90–45 mg/L in hydrodynamic cavitation followed by biological treatment resulting in 30.43–33.34% reduction.

iii) Colour was reduced to 1,800–3,300 hazens in hydrodynamic cavitation followed by biological treatment resulting in 40–47% reduction.
iv) On prior treatment of inlet wastewater using hydrodynamic cavitation, the duration for biological treatment can be reduced, but it still requires a detailed investigation.
v) Using the combination of both the treatment, disposal norms can be achieved.

Overall, the present work assures that combination of biological treatment and hydrodynamic cavitation can effectively be used to achieve disposal norms of treated water which was quiet difficult to do so using anyone of the treatment alone. The investigation also shows that hydrodynamic cavitation suits as pre-treatment technology for CETP wastewater followed by biological treatment.

References

Arrojo, S., Benito, Y. (2008). A theoretical study of hydrodynamic cavitation. Ultrasonics Sonochemistry, 15(3), 203–211. https://doi.org/10.1016/j.ultsonch.2007.03.007.
Badve, M., Gogate, P., Pandit, A., Csoka, L. (2013). Hydrodynamic cavitation as a novel approach for wastewater treatment in wood finishing industry. Separation and Purification Technology, 106, 15–21. https://doi.org/10.1016/j.seppur.2012.12.029.
Chakinala, A.G., Gogate, P.R., Burgess, A.E., Bremner, D.H. (2008). Treatment of industrial wastewater effluents using hydrodynamic cavitation and the advanced Fenton process. Ultrasonics Sonochemistry, 15(1), 49–54. https://doi.org/10.1016/j.ultsonch.2007.01.003.
Chakinala, A.G., Gogate, P.R., Burgess, A.E., Bremner, D.H. (2009). Industrial wastewater treatment using hydrodynamic cavitation and heterogeneous advanced Fenton processing. Chemical Engineering Journal, 152(2–3), 498–502. https://doi.org/10.1016/j.cej.2009.05.018.
Gogate, P.R., Pandit, A.B. (2005). A review and assessment of hydrodynamic cavitation as a technology for the future. Ultrasonics Sonochemistry, 12(1–2 SPEC. ISS.), 21–27. https://doi.org/10.1016/j.ultsonch.2004.03.007.
Gore, M.M., Saharan, V.K., Pinjari, D.V., Chavan, P.V., Pandit, A.B. (2014). Degradation of reactive orange 4 dye using hydrodynamic cavitation based hybrid techniques. Ultrasonics Sonochemistry, 21(3), 1075–1082. https://doi.org/10.1016/j.ultsonch.2013.11.015.
Joshi, R.K., Gogate, P.R. (2012). Degradation of dichlorvos using hydrodynamic cavitation based treatment strategies. Ultrasonics Sonochemistry, 19(3), 532–539. https://doi.org/10.1016/j.ultsonch.2011.11.005.
Mishra, K.P., Gogate, P.R. (2010). Intensification of degradation of Rhodamine B using hydrodynamic cavitation in the presence of additives. Separation and Purification Technology, 75(3), 385–391. https://doi.org/10.1016/j.seppur.2010.09.008.
Ransingh, A., Khambete, A.K. (2017). Hydrodynamic Cavitation, A promising Technique for Pre-Treatment of Common Effluent Treatment Plant (CETP). Influent Wastewater, 2(4), 50–53.
Saharan, V.K., Badve, M.P., Pandit, A.B. (2011). Degradation of Reactive Red 120 dye using hydrodynamic cavitation. Chemical Engineering Journal, 178, 100–107. https://doi.org/10.1016/j.cej.2011.10.018.

Rubina Sahin, Anil Kumar, Rituraj Chandrakar,
Marta Michalska-Domańska, Vikas Dubey

3 The physico-chemical interaction of fluorine with the environment

3.1 Introduction

Due to high level of fluorine intake, approximately 80 million people are affected all over the world. The main source of fluorine in environment are geogenic, such as weathering of fluoride-bearing minerals, dissolution of fluoride ion with main aquifer or other water bodies under suitable condition. From the past two decades, anthropogenic activities are also responsible for fluoride contamination in environment. Industrialization and burning of fossil fuel are common in all. It is the 13th highly abundant element in lithosphere at 600–700 part per million (ppm) by mass or 0.054% by wt% and is not found in elemental form. It exists as fluorine minerals called fluorides. In lithosphere, fluorine occurs most commonly as calcium fluoride (CaF_2) as a main constituent of minerals and other minerals of fluorine are cryolite, fluoroapatite, topaz and so on. Results of many study reports have also proved that the volcanic eruptions release huge amount of gases fluoride and deposition of fluoride-based ashes or lava on the Earth's crust. The mobility of fluoride ion in water resources is not a simple process. There are several geochemical processes evolved under suitable condition. The main processes are dissolution and precipitation of fluoride-bearing mineral with natural aquifer environment. The adsorption/desorption form metal hydroxides, and clay minerals also promote its accumulation in water bodies. Several natural factors such as soil, hydrogeology, rock type, tectonics and climate of terrain which are mostly engrafting fluoride are reported. In this chapter, we classify and describe global fluoride belts (GFB) geologically. The region underlying by metamorphic and crystalline igneous rocks, arid and semiarid condition of sedimentary basin reported maximum fluoride concentration. The critical zones include cratonic area in Asia, Central Africa, African

Rubina Sahin, Department of Basic Science and Humanities, NMDC DAV Polytechnic, Dantewada, Chhattisgarh, India, e-mail: rsm141209@gmail.com
Anil Kumar, Department of Mechanical Engineering, Bhilai Institute of Technology, Durg, Chhattisgarh, India
Rituraj Chandrakar, Department of Mechanical Engineering, NMDC DAV Polytechnic, Dantewada, Chhattisgarh, India
Marta Michalska-Domańska, Institute of Optoelectronics, Military University of Technology, 2 Kaliskiego Str., 00-908 Warsaw, Poland
Vikas Dubey, Department of Physics, Bhilai Institute of Technology, Raipur 393661, India

https://doi.org/10.1515/9783110721355-003

rift valley, South and North America, large sedimentary basin of China and southern America and arid region of China.

3.2 Content of fluorine in the environment

The source of fluorine can be mostly accredited to lithosphere. When Fluorine-bearing minerals and rocks undergo several series of geochemical reactions under the influence of weathering agents. These chemical processes lead to the dissolution of dissolved ions or components of fluorine in biosphere.

3.2.1 Fluorine in atmosphere

In the atmosphere, 0.6 part per billion (ppb) of fluorine are present as organic fluoride compounds and salt spray. A recent report indicates that it is up to 50 ppb in city environments.

In atmosphere, fluorine and its compounds are found naturally by volcanic eruptions, forest fires, marine aerosols, fumaroles and so on. The atmospheric fluorine is generally in the form of inorganic gases or particulate matters and organic compounds of fluorine are also present. Fluorine gas is reported to be widespread in the atmosphere and is one of the phytotoxic elements in atmosphere.

The contribution of fluoride from natural sources such as rock dust and volcanic eruption are relatively less in the atmosphere. Most of these are derived from anthropogenic source which includes fertilizer industries, bricks and ceramic factories, aluminium metallurgy plant, iron and steel production unit, burning of fossil fuels, glass manufacturing and cement industries. These industries releases various types of fluorine-based gases in a large scale such as F_2, hydrogen fluoride (HF), silicon tetrafluoride (SiF_4) and fluosilicic acid (H_2SiF_4). The particulate matters also emit gases such as NaF, CaF_2 and Na_2SiF_6 (Ozsyath, 2009).

Fluorine may be emitted through various processes either as gaseous or particulate matter consisting of either solid fumes or mechanically entrained dust or mists. The formation of fluoride mist by the intermixing of solution of gaseous fluorides as fine water droplets. Similarly, fumes formed by volatilization of metal fluorides may be of fine particle size. The gaseous form mostly contains HF, and SiF_4 and rarely contains boron trifluoride (BF_3). A series of reactions are involved in the volatilization of fluoride, few of which are given in Table 3.1. SiF_4 is commonly formed during the treatment of phosphate rock with H_2SO_4 or phosphoric acid to form superphosphate. The formation of SiF_4 involved series of chemical reactions depending upon the medium, such as acidic or thermal condition, are illustrated in reaction 12 to 19 (Morris et al., 1937; Whynes, 1956). Similarly, H_2SO_4 reacts with CaF_2 (or fluoride-

Table 3.1: Showing the chemical interaction of fluorine compounds with the atmosphere.

Reaction No.	Formation of HF by interaction of fluorine compound with water vapour
1	$CaF_2 + H_2O \rightleftharpoons CaO + 2HF(g)$
2	$2NaF + H_2O \rightleftharpoons Na_2O + 2HF(g)$
3	$2/3\ AlF_3 + H_2O \rightarrow 1/3\ Al_2O_3 + 2HF(g)$
4	$CaF_2 + H_2O + SiO_2 \rightleftharpoons CaSiO_3 + 2HF(g)$
5	$CaF_2 + H_2O + Al_2O_3 \rightleftharpoons Ca(AlO_2)_2 + 2HF(g)$
6	$NaAlF_6 + 2H_2O \rightleftharpoons Na\ AlO_2 + 2NaF + 4HF(g)$
	Formation of volatile metal fluorides
7	$CaF_2 + NaSiO_2 \rightleftharpoons CaSiO_3 + 2NaF(g)$
8	$CaF_2 + K_2SiO_3 \rightleftharpoons CaSiO_3 + 2KF(g)$
9	$CaF_2 + Na_2CO_3 + SiO_2 \rightleftharpoons CaSiO_3 + CO_2(g) + 2NaF(g)$
10	$CaF_2 + 4/3\ Al_2O_3 \rightleftharpoons Ca(AlO_2)_2 + 2/3\ AlF_3(g)$
	Formation of BF₃
11	$6CaF_2 + 5B_2O_3 \rightleftharpoons 4BF_3 + 3Ca_2B_2O_5$
	Formation of SiF₄ in an acidic medium
12	$CaF_2 + H_2SO_4 \rightleftharpoons CaSO_4 + 2HF(g)$
13	$4HF + SiO_2 \rightleftharpoons SiF_4 + 2H_2O(g)$
14	$6HF + SiO_2 \rightleftharpoons H_2SiF_6 + 2H_2O(g)$
15	$H_2SiF_6 \rightleftharpoons SiF_4 + 2HF(g)$
	Formation of SiF₄ in thermal process
16	$Na_2SiF_6 \rightleftharpoons 2NaF + SiF_4(g)$
17	$CaF_2 + 1/2\ SiO_2 \rightleftharpoons Cao + 1/2\ SiF_4(g)$
18	$CaF_2 + 3/2\ SiO_2 \rightleftharpoons CaSiO_3 + 1/2\ SiF_4(g)$
19	$CaF_2 + 1/2\ CaSiO_3 \rightleftharpoons 3/2\ CaO + 1/2\ SiF_4(g)$

bearing minerals such as fluorite, fluorapatite) produces HF. It also reacts with silica (SiO_2) forming subsequent product as given in reaction 4. H_2SiF_6 does not exist as vapour state (Simons, 1950); hence, it is not considered as atmospheric fluoride gases. Table 3.1 gives idea about the formation of HF by the action of water vapour with different metal fluorides at high temperature (>370 °C). For the analysis of fluoride, these reaction are very helpful.

3.2.2 Fluorine in hydrosphere

Seawater contains the average concentration of fluorine which is 1.3 mg/L. In seawater, fluorides are found as MgF^+ ions ranging from 47% to 51% of total fluorine concentration in hydrosphere sediments either as part of marine organisms such as fish bones or shell or precipitated as carbonate fluorapitite (Cappellen and Berner, 1988). Carbonate fluorapitite found to be the most abundant fluoride mineral. Its composition indicates that it is a highly substituted apatite with the chemical formula $Ca_5(PO_4)_2.5(CO_3)0.5$ F and it contains 3.90 wt% of fluorine in seawater.

In water, Fluoride (F^-) is regarded as strong ligand. It can form soluble complexes with polyvalent cations such as Ca^{2+}, Al^{3+}, Mg^{2+} and Fe^{3+}. The formation of complexes between fluoride (F^-) with metal cations depend upon the pH value of water (Nordstorm and Janne, 1977) as well as other interfering species present in water. Few species are boron, beryllium, silicon, vanadium and uranium and rare earth metals. Under acidic condition, free F^- easily combine with metals such as sodium, calcium and magnesium to form compounds. The same pattern of reaction is also reported with neutral species HF. The natural aquatic reservoir contains aqueous specification is known to possess F^- compounds. Table 3.2 showing their thermodynamic data of its formation. In large sedimentary basins located in semiarid regions of the world, such as in the Chacopampean plain in South America, Yellow river basin in China, and the arid and

Table 3.2: List of fluoride species and their thermodynamic data for aqueous reaction.

Element	Ionic species of fluoride in aqueous condition	Chemical reaction involved	Value of log k
Europium	Europium(II)fluoride	$Eu(II) + F^-_{(aq)} \rightleftharpoons EuF^+_{(aq)}$	−1.30
Barium	Barium(II)fluoride	$Ba(II) + F^-_{(aq)} \rightleftharpoons BaF^+_{(aq)}$	−0.18
Lead	Lead(II)fluoride	$Pb(II) + F^-_{(aq)} \rightleftharpoons PbF^+_{(aq)}$	0.81
	Lead(II)difluoride	$Pb(II) + F^-_{(aq)} \rightleftharpoons PbF^+_{(aq)}$	1.62
Copper	Copper(II)fluoride	$Cu(II) + F^-_{(aq)} \rightleftharpoons CuF^+_{(aq)}$	1.20
Manganese	Manganese(II)fluoride	$Mn(II) + F^-_{(aq)} \rightleftharpoons MnF^+_{(aq)}$	1.40
Iron	Ferrous(II)fluoride	$Fe(II) + F^-_{(aq)} \rightleftharpoons FeF^+_{(aq)}$	1.30
	Ferric(II)fluoride	$Fe(III) + F^-_{(aq)} + \rightleftharpoons FeF^{2+}_{(aq)}$	4.10
Samarium	Samarium(III)fluoride	$Fe(III) + F^-_{(aq)} + \rightleftharpoons FeF^{2+}_{(aq)}$	4.35
Tin	Tin(II)fluoride	$Sn(II)F^-_{(aq)} \rightleftharpoons SnF^+_{(aq)}$	4.09
Aluminium	Aluminium(III)fluoride	$Al(III) + F^-_{(aq)} + \rightleftharpoons AlF2 +_{(aq)}$	7.0

Source: Gabriela and Borgnino (2015). Reproduced by permission from The Royal Society of Chemistry.

semiarid regions of southern USA and northern Mexico, groundwater constitutes elevated concentration of F⁻ (Valenzuela, 2006; Currel et al., 2011; Nicolli, 2012).

3.2.3 Fluorine in lithosphere

In lithosphere, fluorine occurs most commonly as CaF_2 as a main constituent of minerals such as cryolite, fluoroapatite and topaz.

Although fluorine and its compounds are abudantly present in the Earth's crust, it is rarely present in the sea and oceans or other water bodies. This may be due to the low water solubility of its salts (fluorides). Among halogens, chlorides have high affinity with water due to which fluorides concentration in sea is 1.3 ppm whereas chlorides concentration is 19,000 mg/L. Most of the world's high F⁻ area are underlain by crystalline igneous and metamorphic rocks (as in part of Sri Lanka, Ghana, Tanzania, Cameroon, India, Scandinavia and the Pampean ranges in South America). The average concentration of fluorine in crystalline rocks can range between 100 mg/kg and 1,000 mg/kg; the data indicate that it is a huge proportion of fluoride-bearing minerals (Fantong, 2010).

Fluorine is 13th most abundant element in lithosphere at 600–700 ppm by mass or 0.054% by weight percentage. It is not found in elemental form and exists as fluorine minerals called fluoride ions – fluorite, also known as fluorspar (CaF_2). It is the most abundant mineral of fluorine. Fluorapatite ($Ca_5(PO_4)_3F$ which is an in inadvertent source of fluoride. Cryolite Na_3AlF_6 used in metallurgy of aluminium. It is one of the fluorine-rich mineral. Topaz is another mineral of fluorine but not as popular as above minerals (Ayoob and Gupta, 2006). It is also reported that during volcanic eruption and geothermal, springs produce trace amount of organofluorines. Recently, another form of fluorides that was discovered named antozonite. A study said that it contains trace quantities, that is 0.04% by weight of diatomic fluorine.

| Fluorite | Fluorapatite | Cryolite |

Figure 3.1: Image of fluorine-bearing mineral.

3.3 Fluorine through natural sources

On the Earth's crust, fluoride solutes are mostly found in metamorphic and igneous rocks. These rocks undergo series of chemical process in the presence of weathering agents, leading to the alteration of rocks and accumulation of dissolved fluoride ion in the aquifer medium. Fluoride minerals are characterized by distinctive crystalline structure and definite chemical composition. Among all some have dominated in chemical composition or other having incorporated as impurities in crystal lattices. Such structure does not affect the physical or chemical properties of the system to some extent or as a whole. The two physical properties such as electronegativity or ionic radius help us to understand the geochemical process between rocks and aquifer. The famous geochemist named Goldschmidt developed four categories of all elements: lithophile, siderophile, atmophile and chacophile (Figure 3.2). Lithophile elements concentrate in the silicate portion of crust and mantle of the Earth and show strong affinity towards silicate phase. The maximum affinity for a metallic liquid phase has been shown by siderophile elements. Atmophile elements are generally volatile and concentrated in the hydrosphere and atmosphere. Chacophile elements have a great affinity for a sulphide liquid phase. It is also minimal on the surface of the earth, and it may be concerted in the core and mantle of the lithosphere.

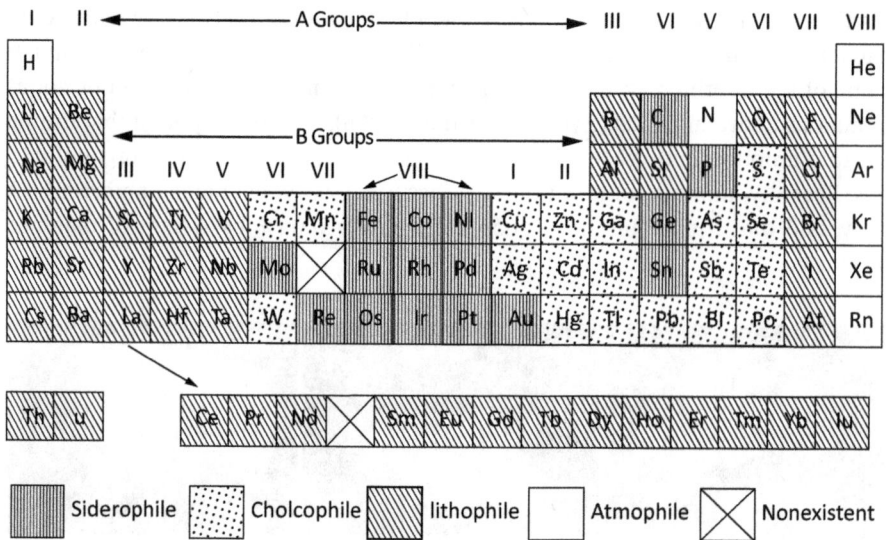

Figure 3.2: Periodic table showing the Goldschmidt's classification of elements.

From Figure 3.1 it is clear that fluorine is lithophile element that has preferentially affinity to silicate phase and concentrate in the silicate portion of the crust and mantle part of the Earth. The geochemical behaviour of fluorine determines its accumulation in the upper continental crust, where its average abundance is 611 mg/kg (Wedepohl, 1995).

As mentioned in Table 3.3, it is clear that large number of fluorine-bearing minerals contain fluorine in their chemical formula as a primary component or incorporates it as an impurity.

Table 3.3: Fluoride-bearing mineral.

Mineral	Chemical formula	Percent (%) fluorine
Fluorapatite	$Ca_{10}(PO_4)_6F_2$	3–4% F
Bastnaesite	(Ce,La) (CO$_3$)F	9.0% F
Cryolite	Na_3AlF_6	45.0% F
Fluorite	(Fluorspar) CaF$_2$	49.0% F
Villianmite	NaF	55.0% F
Sellaite	MgF$_2$	61.0% F

About 1 wt% of fluorine contain muscovite and biotite, while higher percentage contain fluorite (~48 wt%), topaz (~11.5 wt%) and fluorapatite (~3.8 wt%). The fluoride ion (F^-) has similar ionic radius as hydroxyl ion (OH^-). Under suitable condition, replacement of ions may take place in the octahedral sheet of mineral crystal structure as reported in mica (Brigatti and Guggenheim, 2002). In micas lattice structure, the group 17 elements of the periodic table is found at the equivalent level as the apical oxygen of the tetrahedral sheets, where they are bounded to octahedral cation (Figure 3.3). The similar pattern or arrangement also reported in limestone and dolomite (Zidarova, 2010). The fluoride-bearing minerals are also known which contain less amount of fluorine in it such as cryolite (Na_3AlF_6) and topaz ($Al_2SiO_4(OH,F)_2$). Velliamite (NaF) is one of the mineral is almost infinitely soluble, may contribute significantly to the concentration of fluorine in groundwater (Kraynov et al., 1969).

3.3.1 Volcanic sources

Large amount of fluorine enters the environment through volcanic activity and fumarolic gases. During volcanic eruption, lava and magma rise and decompress their volatile species that exsolve into vapour phase. Various types of gases such as H_2O,

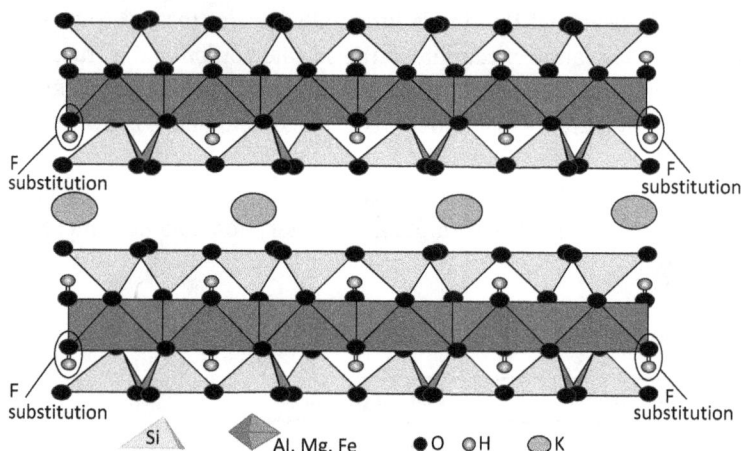

Figure 3.3: Micas crystalline structure showing the substitution of OH⁻ by Fluoride ion (F⁻). Source: Gabriela and Borgnino (2015). Reproduced by permission from The Royal Society of Chemistry.

CO_2, SO_2, NH_3, HF, HCl, H_2S and other minor constituents are emitted by active volcanoes. It is estimated that about 60,000 to 6 million tones of fluorine-based gases are emitted by active volcanoes (Symonds et al., 1988).

The gases emitted through eruption, rapidly react/combine with volcanic plume or ash particles forming thin coating of soluble compounds. These compounds generally consist halide salts or sulphate that are mostly sparingly soluble with fluorine compounds. The compounds are mainly $Ca_5 (PO_4)_3F$, CaF2 and AlF_3 (Deimlie et al., 2007). Apart from that water sources in contact with ash erupted from volcanoes usually contain high concentration of fluorine and its compounds (Ruggeri et al., 2010, Wolff-Boenish et al., 2004). A research paper from Rosi et al., proves that (Calc-) alkaline volcanoes produce relatively fluorine-rich lavas. Such type of volcanoes are found near Japan (island arc) and East Africa (continental rift).

3.3.2 Geothermal sources

Geothermal is another natural sources of fluoride in environment. It is also associated with volcanic activity in active area. Geothermal fluid float over the earth's surface that consists of steam and hot water containing dissolved gases and solutes. They may be discharged from wet and dry-steam and hot-water wells, hot springs and fumaroles. Including a small amount of geothermal water dissolved in magmas which is plagiaristic from gases.

According to Emdunds and Smedley (2013), alkali-chloride solution is a common type of geothermal water having nearly neural pH ($\approx$7). But under different environmental condition, its chemical composition and pH range vary over worldwide. Elements such as boron, lithium, silicon, rubidium, arsenic, caesium, $NH4^+$ and fluorine are commonly dissolved at higher temperature (>250 °C) in geothermal water (Nordstorm and Jenne, 1977). The solubility equilibrium value of fluorite gives idea about the concentration of fluorine in geothermal water and it also depends upon the temperature of fluid. In the western part of USA, F^- concentration upto 50 mg/L was calculated in thermal water collected from Yellowstone Park. In Taiwan, deep geothermal water has an observed concentration of 12 mg/L of fluoride (Deng et al., 2011).

3.3.3 Sea water

The concentration of fluorine in sea water found to be 1.3 mg/L. The marine water contains fluorine in the form of MgF^+ and F^- ions. The percentage of these ions are 47% and 51%, respectively, of the total fluoride concentration in marine water. An aerosol sea spray is produced during removal of fluorine from sea water. Thus, it is precipitated as authigenic minerals mostly carbonate-fluorapatite $Ca_5(PO_4,CO_3)_3F$ and remaining incorporated into marine sediments, either as part of shell or fish bones of marine organisms (Carpenter, 1969).

The most common complex compound of fluorine exists in seawater is in the form of phosphate. The formula of this substituted apatite is $Ca_5(PO_4)_{2.5}(CO_3)_{0.5}F$ that contain ~3.90 wt% F (Van Capllen and Berner, 1988).

3.4 Anthropogenic sources of fluorine

Fluorine reaches atmosphere through natural process is considerably less. Most of their compounds originate through anthropogenic activities. The main sources are fertilizer industries, metallurgy of aluminium and industrial deeds such as ceramic factories, cement works, glass manufacturing and iron and steel industries. The above-mentioned sources release fluorine in the atmosphere in the form of gases such as HF, SiF_4, F_2 and H_2SiF_4 or as particulate matter such as CaF_2, NaF and Na_2SiF_6. Atmospheric CFC is also one of the contributors of fluorine in the atmosphere (Ozsavath, 2009). The concentration of fluorine is not as much as in the ground level. Its concentration through precipitation is reported very low, typically <50 µg/L.

Hydrogen fluoride (HF) is one of the fluorine compounds present in the atmosphere. This is emitted as reduction of aluminium with small amount of perfluoromethane (CF_4) and sulphur hexafluoride (SF_6) and is released from magnesium

Table 3.4: Showing extent of emission of fluoride gases and particulate from various industries.

Process	Equipment of the process	Form of fluorine	Concentration of fluorine $\mu g/m^3$
Steel Manufacture	Open hearth furnace	HF	2.2883519×10^6
Steel Manufacture	Open hearth furnace	Particulates	1.8306815×10^4
Aluminium Alloy processing	Open hearth furnace	Gaseous	$1.1441759 \times 10^4 - 2.7460222 \times 10^4$
Manufacture of superphosphate	Continuous dens	SiF_4	$2.28835190 \times 10^6 - 5.2632093 \times 10^7$
Deflorinating phosphate rock (molten)	Shaft furnace	HF	$3.2036926 \times 10^6 - 6.40738532 \times 10^6$
Nodulizing phosphate rock	Rotary Klin	HF	2.4027694×10^6
Brick works	Klin		5.9497119×10^5
Aluminium melting	Open hearth reverberatory furnace	Particulates	$2.517187 \times 10^5 - 1.8306815 \times 10^6$

industry. The very toxic gas sulphuryl fluoride (SO_2F_2) used as fumigant in ships, warehouses and stores.

Anthropogenic sources contributing fluoride are mainly through mining activities, phosphate fertilizer effluents and runoff from Earth's surface that reaches the water bodies. Globally, fluoride in groundwater is mostly due to geogenic in nature.

3.5 Process involved in geochemistry of fluorine

The occurrence, distribution and movement of fluorine in aquifer depend on the interaction of a number of geochemical processes that determines its elimination or release into the solution. Environment condition such as chemical composition of the system, that is rock types, pH values and kinetics of chemical reaction during all geothermal processes (Figure 3.4).

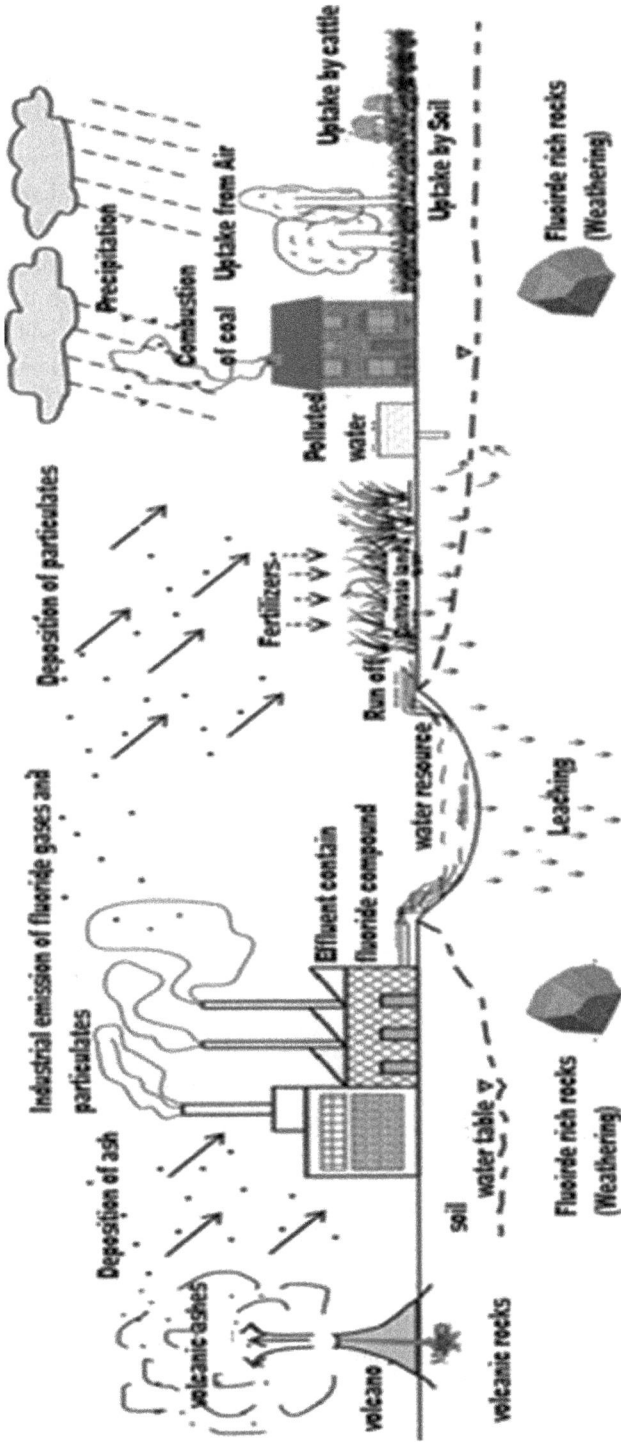

Figure 3.4: Schematic diagram showing natural and anthropogenic sources of fluoride in the environment.

3.5.1 Process of adsorption

Environment has unique properties of self supporting and self-controlling for which colloids play vital role in the controlling, cycling and transport of pollutant among different sphere. It may hold on to or release ions through the process of adsorption and desorption. Adsorption is a process of transfer of adsorbate ions (adsorbate) from the aqueous phase to the solid adsorbent phase through different mechanisms such as surface precipitation and the adsorption based on physical and chemical attraction. The process of adsorption can also be defined as a set of reaction between the ions of solutes and functional groups attached onto the surface of solid adsorbent. These are number of natural colloidal components to promote the rate of wide range of variety of functional groups lying on the surface that are formed during adsorption process of solid–water interaction.

The metal ions behave as Lewis acid because their coordination number decrease due to adsorption process on the surface layer. In the water–solid interaction, the surface of metal ion may first try to coordinate with water molecule (Figure 3.5(i)) and the dissociative chemisorptions process of water leads to the formation of hydroxyl group (OH) on the surface (Figure 3.5(ii)). Based on this condition, two possibilities are there – First possibility are the exchange of ion in aqueous solution and another possibility are deprotonation. In both the processes, functional groups are highly reactive (Stumm, 1992).

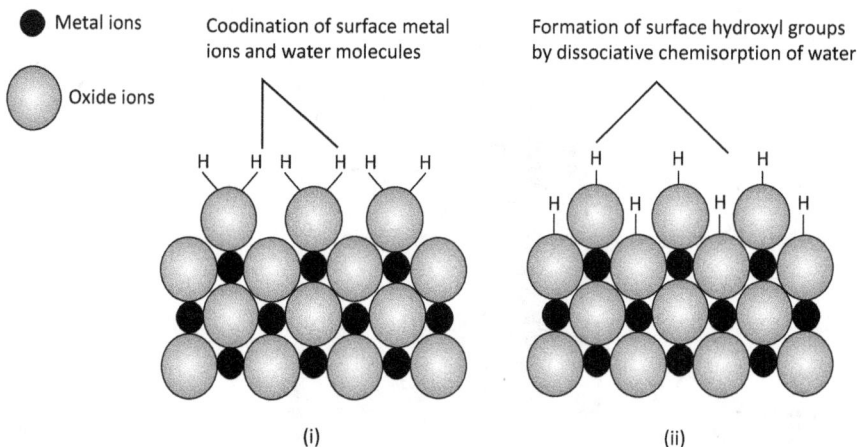

Figure 3.5: The interface of solid–water of hydroxides clays minerals, oxides of metal, amorphous silicates and oxy-hydroxides contain functional groups on their surface.

The specific adsorption occurs in inner sphere complexes. The short range bonding between the solute complex and oxygen surface are formed during complexation process. This type of interaction is also known as chemisorptions. The non-specific adsorption seen in outer sphere complexes. In this type of adsorption, the process of complexation takes place between aqueous ions and the functional group which are present on the surface through electrostatic interaction (Figure 3.6). Such interaction is known as physisorption. It is reported that, the chemical attraction of an ion on the surface can make the electrostatic attraction ineffective. It means that, strong chemical interaction involved the adsorption of negatively charged ion onto the positively charged surface.

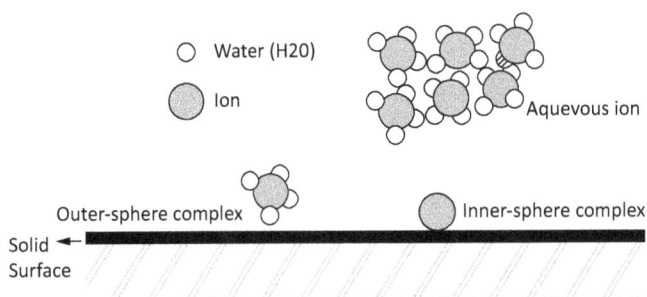

Figure 3.6: Showing outer surface complexes and inner surface complexes on solid surfaces.

Ions which are attached by electrostatic forces are easily displaced by ions with similar charge or ionic radius, and thus it is termed as exchangeable. The exchangeable ions play vital role to maintain the nutrients level in plants. But in case of environmental pollutants, it is not strong enough to put out of action. The chemically engaged ions show very strong bonding with solute (solid) and are considered as irreversible. Thus ions which are chemically bonded have fewer tendencies to release as compared to the ions which are held by physical force, that is, electrostatic forces.

To understand the characterization and interaction of fluorine with mineral surfaces, several studies are carried out. The adsorption of fluoride ion (F^-) onto solid surface of adsorbent normally involves three important steps:

i) Transport or diffusion of F^- to the outer surface of the adsorbent from bulk aqueous solution across the boundary layer sounding the solid adsorbent particle. This is called external mass transfer.

ii) F^- adsorbed on to surface of solid particle.

iii) Depending upon the chemistry of solid adsorbent, the adsorbed F^- may be exchanged within the structural unit of adsorbent from inside.

Or the adsorbed F^- are transferred/diffused to the internal surfaces for porous materials (commonly known as intra-particle diffusion).

The adsorption and desorption of fluorine over the mineral surface take place under suitable aquatic environment such as pH value, stability of adsorbent, adsorption capacity in dilution condition, loading capacity of ions in the presence of other anions and regeneration ability. Mostly clays, metal (hydroxides) and carbonates acquire such type of mineral surfaces which develop the environmen for the adsorption/desorption of the fluoride ion (F^-) onto the surface during adsorption phenomenon.

3.5.2 Adsorption of fluorine onto calcite and metal hydroxides

Calcite ($CaCO_3$) is an chief mineral that penetrate the surface of soils and sediments, where it is commonly present in cement or dendrite grain. The calcite consists of calcium carbonate ($CaCO_3$) in the range of 90–97%, magnesium oxide (MgO) 1.5–2.5 and silica 1–3.5%. In wastewater treatment technologies, this mineral is used as large-scale adsorbent because it has strong affinity of fluoride ion (F^-) for calcite sites. The model proposed by Stipps explained that on the surface of calcite, calcium and carbonate acquire partial charge, that is > $Ca^{\partial+}$ and > $CO_3^{\partial-}$ along any cleavage, where symbol > represent sedge of calcite bulk surface. On hydrolysis, these charges imbalance and the value decreases with the species that have strong affinity to react with any other ions in the solution. At neutral pH (7) or higher value, ions like OH^- and CO_3^{2-} dominate producing > $CaCO_3^-$ and CO_3^+ sorption sites. On decreasing pH value of the system, positive surface sites increases, leading to an increase in the adsorption of fluoride (F^-) ion on the surface sites (Turner et al., 2005a & 2005b. Thus, positive sorption occurs via the following reaction:

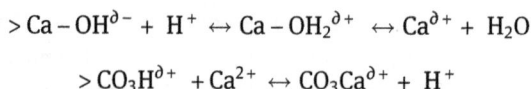

$$> Ca - OH^{\partial-} + H^+ \leftrightarrow Ca - OH_2^{\partial+} \leftrightarrow Ca^{\partial+} + H_2O$$

$$> CO_3H^{\partial+} + Ca^{2+} \leftrightarrow CO_3Ca^{\partial+} + H^+$$

The efficiency of fluoride ion (F^-) uptake of calcite through adsorption technique is calculated as 75.6% on pH range 6.7–7.4. The F^- uptake onto the surface of calcite follows the pseudo-second order equation.

The adsorption of F^- ion onto various metal hydroxides has been studied. Among all the studies, adsorption on iron hydroxides has been investigated extensively (Hiemstra and Riemsdijik, 2000; Sahin et al, 2016).The kinetics of F^- adsorption onto adsorbent surface is relatively fast, between 10 min and 10 h, depending upon oxide crystalline. The pH range between 4 and 8 is found to be the optimal value for maximum adsorption. Mostly acidic environment favours the process of adsorption on maximum extent. During the process, F^- interacts with surface group by inner sphere surface complexes, and the replacement of hydroxyl group (OH^-) by the F^- takes place and simultaneously liberate hydroxyl ion (OH^-) into the solution as shown in the following reaction:

$$\equiv FeOH^{-1/2} + F^- \rightleftharpoons \equiv FeF^{-1/2} + OH^-$$

Farrah (1987) was the first researcher to investigate the interaction of fluoride (F^-) with aluminium hydroxide at optimum pH range of 3 to 8.

In the study he investigated that under strong acidic condition (pH > 4), the oxides dissolve gradually with the formation of Aluminium–Fluorine (Al–F) complexes, while in the pH range between 5.5 and 6.5 maximum adsorption of fluoride (F^-) was reported, that is 170 mg/gm (Farrah et al., 1987).

In case of aluminium oxides, adsorption of fluoride (F^-) is low and calculated its value is 0.2–0.4 mg/gm. To enhance its adsorption efficiency, it must be activated by heating or some other reagents.

3.5.3 Adsorption onto clay minerals and soil

Electrically, the soil is found to be negatively charged, which make these material not suitable for removal of anions. But such type of condition is transformed in case of sedimentary deposits where soils or clays are typically connected with metal hydroxide coating mostly with iron hydroxide (Zhung and Yu, 2002). Thus, such type of surface coating develop the possibility of adsorption of fluoride (F^-) on it. In the soil system, the formation of Al–F complexes also increases the adsorption capacity (Harrington et al., 2003) (Zhu et al., 2004). It indicates that the complexation of fluoride (F^-) with surficial aluminium and iron via exchange of ligand with sacrificial hydroxyl group (OH^-). in the presence of water. The mechanism of adsorption on to clay minerals have been observed in the presence of water molecule. It means that adsorption mechanism is dominated in aqueous condition (Bia et al., 2012).

In soils and sedimentary deposits, the competitive adsorption is a common mechanism (Zhu et al., 2007). Here, fluoride (F^-) compete with other anions such as chloride (Cl^-), sulphate (SO_4^{2-}), phosphate (PO_4^{3-}), carbonate (CO_3^{2-}) and arsenate (As^{2-}) (Bia et al., 2012) over the surface sites in the aquifer environment.

3.5.4 Dissolution and precipitation

In a closed system of the environment, the composition and quantity of a mineral that precipitates or dissolves can be explained in terms of chemical kinetics and thermodynamics as influenced by the surface morphology of species which are dissolving or precipitating as shown in Figure 3.7. It is quite impossible to completely understand the pathways and processes of dissolution and precipitation, assuming the interaction among kinetics, thermodynamic and surface morphology.

The chemical thermodynamic helps to predict the standard free energy change of chemical reaction (ΔG°_r) from mineral and solution specification and also to determine the composition of mineral and soil solution. If the system attains the equilibrium, it means that ΔG°_r reaches a minimum. If it is in non-equilibrium condition, then system is metastable product where it opposes dissolution or precipitation process. And if irreversible reaction occurs, that means rock component dissolves completely (Figure 3.7). The rate of dissolution and precipitation may be very small because the driving force (i.e energy change of the system) is too small. Whether the chemical reaction is possible or not, only information is given by thermodynamic whereas kinetics provide information about the time required for such changes or transformation (2 in Figure 3.7). The chemical kinetics also takes consideration of transport of ions in solution, crystal growth, rate of nucleation, reaction rate of solution and dissolution. During the chemical kinetics and thermodynamic, the energy of the system changed, and this can be customized by the surface morphology of the mineral such as structure, thickness, composition and surface area. During chemical reaction, formed soluble products envisaged by thermodynamic can persuade all the aspects of surface morphology (3 in Figure 3.7). For example, nucleation and crystal growth could produce new particles on a surface. On the other hand, the processes of dissolution could alter the surface by producing leached layers, or crystal ripening could create crystals of smaller surface area. In turn, surface morphology can affect the release or merging of solution components which change the free energy of solution, and hence mineral reaction pathways may be transformed (4 in Figure 3.7). Kinetic factors can affect the surface morphology (e.g., incongruent dissolution creates "leached layers" at a surface; 5 in Figure 3.7) just as much as surface morphology will utter the speed of dissolution and precipitation (6 in Figure 3.7).

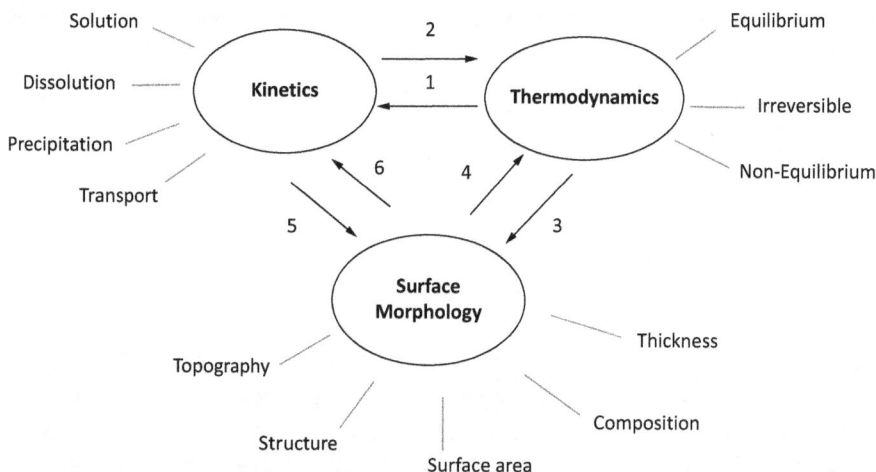

Figure 3.7: Three components influencing dissolution and precipitation of fluoride-bearing minerals.

Fluorite and fluorapatite are common minerals that may precipitate from saturated solution in the aquifer environment. This process depends upon the alkalinity of the media solubility constant of these minerals or presence of other carbonate fluorapatite.

In the natural environment, the common fluoride-bearing minerals are micas, fluorite, amphiboles, apatite or in lesser extent cryolite and topaz. As per solubility constant values, these minerals dissolve in aquatic medium. According to Table 3.5, it is clear that fluorite and cryolite are much more squabble then micas and fluorapatite in the pH range between 4 and 9 in natural water. The solubility is almost same across the entire range of this pH. Several studies indicate that solubility of micas and fluorapitie are vastly depend on the value of pH. The sensitive range of pH found to be between pH 2.5 and 12.5. The saturation of aquifer with respect to fluorite and calcite has computed with the help of the following reaction mechanism:

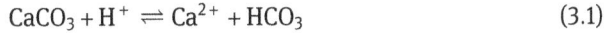

$$CaCO_3 + H^+ \rightleftharpoons Ca^{2+} + HCO_3 \tag{3.1}$$

$$K_{cal} = a_{Ca2+} \cdot a_{HCO3-}/a_{H+} = 0.97 \times 10^2 \text{ (Hem, 1970)}$$

where a = activity of the ions in moles/L

Table 3.5: List of the fluoride-bearing minerals and their solubility constant at 25 °C.

S. no.	Fluoride mineral and their chemical formula	Value of log k_{ps}
1	Fluorite (CaF_2)	−10.0371
2	Fluorapatite ($Ca_5(PO_4)_3F$	−24.9941
3	Carbonate-fluorapatite ($Ca_{9.316}Na_{0.36}Mg_{0.144}(PO_4)_{4.8}(CO_3)_{1.2}F_{2.48}$	−114.40
4	Cryolite (Na_3AlF_6)	−53.40

Similarly, to express the solubility of fluorite the following chemical reactions are considered:

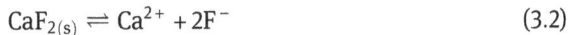

$$CaF_{2(s)} \rightleftharpoons Ca^{2+} + 2F^- \tag{3.2}$$

$$K_{fluorite} = a_{ca2+} \cdot (a_F^-)^2 = 10^{-10.5} \text{ (Smyshlyaevx and Edeleva 1962)}$$

In order to calculate the thermodynamic equilibrium in aquifer which are in contact with both calcite and fluorite solid phase, a combined mass law equation relating both the solute species can be written as:

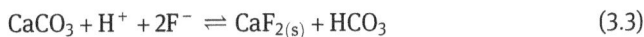

$$CaCO_3 + H^+ + 2F^- \rightleftharpoons CaF_{2(s)} + HCO_3 \tag{3.3}$$

$$K_{cal-fluorite} = a_{HCO3-}/a_{H+} \cdot (a_F^-)^2$$

The value for above constant is calculated by Smyshlyaev and Edeleva 3.6×10^{12}. The above relation sustain if the pH of the system remains constant. But any increase and decrease in HCO_3^- concentration/activity of F^- since $K_{cal\text{-}fluorite}$ is constant, which indicates a +ve correlation between these two variables (Handa, 1975).

A considerable volume of literature explained that dissolution of fluorite (CaF_2) and fluorapatite pulled by calcite precipitation. The possible reaction mechanism for the fluoride (F^-) dissolution are given below:

Mechanism (1), the fluoride (F^-) released by the dissolution of fluorapatite

$$Ca_5(PO_4)_3F + 3H^+ \rightleftharpoons 5Ca^{2+} + 3HPO_4^{2-} + F^- \tag{3.4}$$

Dissolution of fluorapatite for is decoupled from precipitation of calcite.

Mechanism (2), the fluorapatite dissolution and calcite precipitation can be explained as follows:

$$Ca^{2+} + CO_3^{2-} \rightleftharpoons CaCO_3 \tag{3.5}$$

$$Ca_5(PO_4)_3F + 3H^+ \rightleftharpoons 5Ca^{2+} + 3HPO_4^{2-} + F^- \tag{3.6}$$

Here, it is clear that precipitation of calcite (reaction v) and dissolution of fluorapatite (reaction vi) are chemically coupled by Ca^{2+}; the precipitation of calcite is triggering the dissolution of fluorapatite by reducing the calcium activity of the solution. The combined mass of above chemical equation relating the solute species in which aquifer is in contact with both fluorapatite and calcite can be written as:

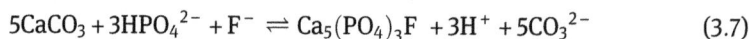

$$5CaCO_3 + 3HPO_4^{2-} + F^- \rightleftharpoons Ca_5(PO_4)_3F + 3H^+ + 5CO_3^{2-} \tag{3.7}$$

The equilibrium constant of this reaction (vii)

$$K_{cal\text{-}fluorapatite} = (a_{H+})^3 (a_{CO3}^{2-})^5 / (a_{HPO4}^{2-})^3 \cdot (a_f^-)$$

It resembles the equilibrium relationship between fluorite and calcite (Handa, 1975). It provides natural and interdependent control on concentration of Ca^{2+}, F^-, CO_3^{2-}, PO_4^{2-} and dissolved organic carbon in the solution.

References

Aoba, T., Fejerskov, O. (2002). Dental fluorosis: Chemistry and biology. Critical Reviews in Oral Biology and Medicine, 13, 155–170.

Ayoob, S., Gupta, A.K. (2006). Fluoride in drinking water: A review on the status and stress effects. Critical Reviews in Environmental Science and Technology, 36, 433–4 87.

Banerjee, A. (2015). Groundwater fluoride contamination: A reappraisal. Geoscience Frontiers, 6, 277–284.

Berger, T., Peltola, P., Drake, H., Åström, M. (2012). Impact of a fluorine-rich granite intrusion on
    levels and distribution of fluoride in a small boreal catchment. Aquatic Geochemistry, 18,
    77–94.
Bernal, N.F., Gleeson, S.A., Dean, A.S., Liu, X.-M., Hoskin, P. (2014). The source of halogens in
    geothermal fluids from the Taupo Volcanic Zone, North Island, New Zealand. Geochimica et
    Cosmochimica Acta, 126, 265–283.
Bia, G., De Pauli, C.P., Borgnino, L. (2012). The role of Fe(III) modified montmorillonite on fluoride
    mobility: Adsorption experiments and competition with phosphate. Journal of Environmental
    Management, 100, 1–9.
Borgnino, L., Garcia, M.G., Bia, G., Stupar, Y., Coustumer, P.L., Depetris, P.J. (2013). Mechanisms of
    fluoride release in sediments of Argentina's central region. Science of the Total Environment,
    443, 245–255.
Brigatti, M.F., Guggenheim, S., Mica crystal chemistry and the influence of pressure, temperature,
    and solid solution on atomistic models, Micas: Crystal Chemistry & Metamorphic Petrology,
    Mottana, A., Sassi, F.P., Thompson, J.B. Jr., Guggenheim, S., Mineralogical Society of America,
    2002, pp. 1–98.
Burt, B.A., Eklund, S.A. (1992). Dentistry, Dental Practice and the Community, WB Saunders,
    Philadelphia, 147.
Carpenter, R. (1969). Factors controlling the marine geochemistry of fluorine. Geochimica et
    Cosmochimica Acta, 33, 1156–1167.
Chae, G.T., Yun, S.T., Kwon, M.J., Kim, Y.S., Mayer, B. (2006). Batch dissolution of granite and
    biotite in water: Implication for fluorine geochemistry in groundwater. Geochemical Journal,
    40, 95–102.
Chae, G.T., Yun, S.T., Mayer, B., Kim, K.H., Kim, S.Y., Kwon, J.S., Kim, K., Koh, Y.-K. (2007). Fluorine
    geochemistry in bedrock groundwater of South Korea. Science of the Total Environment, 385,
    272–283.
Chaïrat, C.J., Schott, E.H., Oelkers ., J.-E.L., Harouiya, N. (2007). Kinetics and mechanism of natural
    fluorapatite dissolution at 25 °C and pH from 3 to 12. Geochimica et Cosmochimica Acta, 71,
    5901–5912.
Pan, Y., Fleet, M.E. (2002). Compositions of the Apatite-group Minerals: Substitution Mechanisms
    and Controlling Factors. Phosphates, Kohn, M. L., Rakovan, J. and Hughes, J. M., Mineralogical
    Society of America, 13–50.
Cronin, S.J., Neall, V.E., Lecointre, J.A., Hedley, M.J., Loganathan, P. (2003). Environmental hazards
    of fluoride in volcanic ash: A case study from Ruapehu volcano, New Zealand. Journal of
    Volcanology and Geothermal Research, 121, 271–291.
Currel, M., Cartwright, I., Raveggi, M., Han, S. (2011). Controls on elevated fluoride and arsenic
    concentrations in groundwater from the Yuncheng Basin, China. Applied Geochemistry, 26,
    540–552.
Delmelle, P., Lambert, M., Dufrene., Y., Gerin, P.,., Oskarsson, N. (2007). Gas/aerosol–ash
    interaction in volcanic plumes: New insights from surface analyses of fine ash particles. Earth
    and Planetary Science Letters, 259, 159–170.
Deng, Y., Nordstrom, D.K., McCleskey, R.B. (2011). Fluoride geochemistry of thermal waters in
    Yellowstone National Park: I. Aqueous fluoride speciation. Geochimica et Cosmochimica Acta,
    75, 4476–4489.
Edmunds, W.M., Smedley, P.L. (2013). Fluoride in natural waters, Essentials of Medical Geology:
    Revised Edition, Selinus, O.British Geological Survey, 311–336.
Fantong, W.Y., Satake, H., Ayonghe, S.N., Suh, E.C., Adelana, S.M.A., Fantong, E.B.S., Banseka,
    H.S., Gwanfogbe, C.D., Woincham, L.N., Uehara, Y., Zhang, J. (2010). Geochemical provenance
    and spatial distribution of fluoride in groundwater of Mayo Tsanaga River Basin, Far North

Region, Cameroon: Implications for incidence of fluorosis and optimal consumption dose. Environmental Geochemistry and Health, 32, 147–163.

Farrah, H., Slavek, J., Pickering, W.F. (1985). Fluoride sorption by soil components, calcium carbonate, humic acid, manganese dioxide and silica. The Australian Journal of Soil Research, 23, 429–439.

Farrah, H., Slavek, J., Pickering, W.F. (1987). Fluoride interactions with hydrous aluminum oxides and alumina. Australian Journal of Soil Research, 25, 55–69.

Fawell, J., Bailey, K., Chilton, J., Dahi, E. (2006). Fluoride in Drinking-Water, World Health Organization, Cornwall, UK, 144.

Fenter, P., Gessbuhler., P., Srajer., G., Sorenson, L.B., Starchio, N.C. (2000). Surface specification of calcite observed in situ by high-resolution X-ray reflectivity. Geochimica et Cosmochimica Acta, 64, 1221–1228.

García, M.G., Borgnino, L. Chapter 1: Fluoridein the Context of the Environment, in Fluorine: Chemistry, Analysis, Function and Effects, 2015, 3–21

García, M.G., Lecomte, K.L., Stupar, Y., Formica, S.M., Barrionuevo, M., Vesco, M., Gallará, R., Ponce, R. (2012). Geochemistry and health aspects of F-rich montainous streams and groundwaters from Sierras Chicas de Córdoba, Argentina. Environmental Earth Sciences, 65, 535–545.

Guo, Q. (2012). Hydrogeochemistry of high-temperature geothermal systems in China: A review. Applied Geochemistry, 27, 1887–1898.

Guo, Q., Wang, Y., Ma, T., Ma, R. (2007). Geochemical processes controlling the elevated fluoride concentrations in groundwaters of the Taiyuan Basin, Northern China. Journal of Geochemical Exploration, 93, 1–12.

Handa, B.K. (1974). Methods of collection and analysis of water samples and interpretation of water analysis, Data.Govt. of India, 365.

Handa, B.K. (1975). Geochemistry and genesis of fluoride-containing groundwaters in India. Groundwater, 13, 275–281.

Harrington, L.F., Cooper, E.M., Vasudevan, D.J. (2003). Fluoride sorption and associated aluminum release in variable charge soils. Journal of Colloid and Interface Science, 267, 302–313.

He, J., An, Y., Zhang, F. (2013). Geochemical characteristics and fluoride distribution in the groundwater of the Zhangye Basin in Northwestern China. Journal of Geochemical Exploration, (in press).

Hiemstra, T., Van Riemsdijk, W.H. (2000). Fluoride adsorption on goethite in relation to different types of surface sites. Journal of Colloid and Interface Science, 225, 94–104.

Hudson-Edwards, K.A., Archer, J. (2012). Geochemistry of As-, F- and B-bearing waters in and around San Antonio de los Cobres, Argentina, and implications for drinking and irrigation water quality. Journal of Geochemical Exploration, 112, 276–284.

Hy Em, J.D. (1970, 1473). Study and interpretation of the chemical characteristics of natural water, U.S. Geological Survey, Water supply paper.

IBID, 85th Annual report, 1949.

IBID, 86th Annual report, 1950.

IBID 87th Annual report. 1951.

IBID 89th Annual report, 1953.

IBID 90th Annual report, 1954.

Ivanova, T.I., Frank-Kamenetskaya, O.V., Kol'tsov, V., Ugolkov, L. (2001). Crystal structure of calcium-deficient carbonated hydroxyapatite. Thermal decomposition. Journal of Solid State Chemistry, 160, 340–349.

Kraynov, S.R., Merkov, A.N., Petrova, N.G., Baturinskaya, I.V., Zharikova, V.M. (1969). Highly alkaline (pH 12) fluosilicate waters in the deeper zone of the Lovozero Massif. Geochemistry International, 6, 635–640, Search.

Liu, G., Zheng, L., Qi, C., Zhang, Y. (2007). Environmental geochemistry and health of fluorine in Chinese coals. Environmental Geology, 52, 1307–1313.

Nicolli, H.B., Bundschuh, J., Blanco, M.D.C., Tujchneider, O.C., Panarello, H.O., Dapeña, C., Rusansky, J.E. (2012). Arsenic and associated trace-elements in groundwater from the Chaco-Pampean plain, Argentina: Results from 100 years of research. Science of the Total Environment, 429, 36–56.

Nordstrom, D.K., Jenne, E.A. (1977). Fluorite solubility equilibria in selected geothermal waters. Geochimica et Cosmochimica Acta, 41, 175–188.

Ozsvath, D.L. (2009). Fluoride and environmental health: A review. Reviews in Environmental Science and Biotechnology, 8, 59–79.

Padhi, S., Muralidh, D. (2012). Fluoride occurrence and mobilization in geo-environment of semi-arid Granite watershed in southern peninsular India. Environmental Earth Sciences, 66, 471–479.

Pickering, W.F. (1985). The mobility of soluable fluoride in soils. Environmental Pollution, 9, 281–308.

Rango, T., Bianchini, G., Beccaluva, L., Ayenew, T., Colombani, N. (2009). Hydrogeochemical study in the Main Ethiopian Rift: New insights to the source and enrichment mechanism of fluoride. Environmental Geology, 58, 109–118.

Reddy, D.V., Nagabhushanam, P., Sukhija, B.S., Reddy, A.G.S., Smedley, P.L. (2010). Fluoride dynamics in the granitic aquifer of the Wailapally watershed, Nalgonda district, India. Chemical Geology, 269, 278–289.

Reyes-Gómez, V.M., Alarcón-Herrera, M.T., Gutiérrez, M., López, D.N. (2013). Fluoride and arsenic in an alluvial aquifer system in Chihuahua, Mexico: Contaminant levels, potential sources, and co-occurrence. Water Air and Soil Pollution, 224, 1433–1448.

Ritchie, G.S.P. (1989). Role of dissolution and precipitation of minerals in controlling soluble aluminium in acidic soils, Academic Press. Sanoiego, 1–37.

Rosi, M., Papale, P., Lupi, L., Stoppato, M. (2003,). Volcanoes, Firefly Books Ltd., Spain.

Ruggeri, F., Saavedra, J., Fernandez-Turiel, J.L., Gimeno, D., Garcia-Valles, M. (2010). Environmental geochemistry of ancient volcanic ashes. Journal of Hazardous Materials, 183, 353–365.

Sahin, R. (2017). Inferring the chemical parameters for the dissolution of fluoride in groundwater of Bastar zone, Chhattisgarh India. Research Journal of Chemical Sciences, 7, 1–6.

Sahin, R., Tapadia, K., Sharma, A. (2016). Kinetic and isotherm studies on adsorption of fluoride by limonite with batch techniques. Journal of Environmental Biology, 37, 1–8.

Scanlon, B.R., Nicot, J.P., Reedy, R.C., Kurtzman, D., Mukherjee, A., Nordstrom, D.K. (2009). Elevated naturally occurring arsenic in a semiarid oxidizing system, Southern High Plains aquifer, Texas, USA. Applied Geochemistry, 24, 2061–2071.

Semrau, K.T. (1957). Emission of fluorides from Industries processes- A review. Journal of the Air pollution Control, 7, 92–108.

Shrott, N.E., Pandit, C.G., Raghvachari, T.N.S. (1937). Endemic fluorosis in Nellore district of South India. The Indian Medical Gazette, 72.

Simons, J.H. (1950). Fluorine chemistry, ed 1, Academic Press Inc, New York.

Smyshlyaev, S.I., Edeleva, N.P. (1962). Determination of the solubility of minerals, I Solubility product of fluorite. Izv. Vysshhikh Uchebn. Zavedenii Khimi Khim Teckhnologiya, 5.

Stumm, W. (1992). Chemistry of the Solid-Water Interface, Wiley Interscience Publication.

Sujana, M.G., Anand, S. (2010). Iron and aluminium based mixed hydroxides: A novel sorbent for fluoride removal from aqueous solutions. Applied Surface Science, 256, 6956–6962.

Symonds, R.B., Rose, W.I., Reed, M.H. (1988). Contribution of Cl– and F– bearing gases to the atmosphere by volcanoes. Nature, 334, 415–419.

Turner, B.D., Binning, P., Stipp, S.L.S. (2005). Fluoride removal by calcite: evidence for fluoride precipitation and surface adsorption. Environmental Science & Technology, 39, 9561–9568.

Turner, B.D., Binning, P., Stipp, S.L.S. (2005). Fluoride removal by calcite, Evidence for fluorite precipitation and surface adsorption. Environmental Science & Technology, 39, 9561–9568.

Valenzuela Vásquez, L., Ramírez Hernández, J., López, J.R., Sol Uribe, A., Mancilla, O.L. (2006). The origin of fluoride in groundwater supply to Hermosillo City, Sonora, México. Environmental Geology, 51, 17–27.

Van Cappellen, P., Berner, R. (1988). A mathematical model for the early diagenesis of phosphorous and fluorine in marine sediments: Apatite precipitation. American Journal of Science, 288, 289–333.

Van Cappellen, P., Charlet, L., Wersin, P. (1993). A surface complexation model of the carbonate mineral-aqueous solution interface. Geochimica et Cosmochimica Acta, 57, 3505–3518.

Vikas, C., Kushwaha, R., Ahmad, W., Prasannakumar, V., Reghunath, R. (2013). Genesis and geochemistry of high fluoride bearing groundwater from a semi-arid terrain of NW India. Environmental Earth Sciences, 68, 289–305.

Vinson, D.S., McIntosh, J.C., Dwyer, G.S., Vengosh, A. (2011). Arsenic and other oxyanion-forming trace elements in an alluvial basin aquifer: Evaluating sources and mobilization by isotopic tracers (Sr, B, S, O, H, Ra). Applied Geochemistry, 26, 1364–1376.

Wedepohl, K.H. (1995). The composition of the continental crust. Geochimica et Cosmochimica Acta, 59, 1217–1232.

Wolff-Boenisch, D., Gislason, S.R., Oelkers, E.H. (2004). The effect of fluoride on the dissolution rates of natural glasses at pH 4 and 25 °C. Geochimica et Cosmochimica Acta, 68, 4571–4582.

Zhu, M.-X., Ding, K.-Y., Jiang, X., Wang, -H.-H. (2007). Investigation on co-sorption and desorption of fluoride and phosphate in a red soil of China. Water Air and Soil Pollution, 183, 455–465.

Zhu, M.-X., Jiang, X., Ji, G.-L. (2004). Interactions between variable-charge soils and acidic solutions containing fluoride: An investigation using repetitive extractions. Journal of Colloid and Interface Science, 276, 159–166.

Zhuang, J.I.E., Yu, G.-R. (2002). Effects of surface coatings on electrochemical properties and contaminant sorption of clay minerals. Chemosphere, 49, 619–628.

Zidarova, B. (2010). Hydrothermal fluorite-forming processes in the Mikhalkovo deposit (Central Rhodopes, Bulgaria) field observation and experimental confirmation. Neues Jahrbuch für Mineralogie, 187, 133–157.

Ghanshyam Shakar, Bhumika Das

4 The effect of chemical waste produced by industrial area near Raipur (Chhattisgarh) on the quality of drinking water

Abstract: Water is the main component on earth for the life of people. Groundwater is the most perfect wellspring of water accessible to satisfy our ordinary needs. This is the reason why the dependence on groundwater has ascended so much that it has added to groundwater overexploitation. Regarding groundwater, a few urban communities in India have just arrived at zero levels. Its yield is declining because of overexploitation and absence of groundwater revives. Notwithstanding over misuse, people have additionally changed the characteristic groundwater energize framework by building homes, street organizations, production lines and different administrations. Modern waste is not sufficiently discarded by the industrial facilities and is generally flown through the open land and stream/drain-pipe channels, which has led to surface water contamination. Both overexploitation and mechanical waste would be destroying in the coming days. In this paper, authors analysed the harmful effects of the industrial wastes in nearby places of Raipur (CG). Eleven different samples of water have been collected from the industrial area, Siltara, which is contaminated by the industrial wastes. We have analysed the water sample by measuring pH, TDS, colour, DO and so on. It was found that the water is contaminated with the higher risk levels and different measures have to be taken to purify it.

4.1 Introduction

The most basic worldwide ecological, social and political issue (APHA, 21st version) is the quality, amount and accessibility of drinking water. Ground water is one of the critical consumable water and, in view of its relative objectivity, it is considerably very hard to treat this groundwater if contaminated (BIS, 2012). Despite influencing water quality, harmful material additionally compromises human being, financial development and social security (Punamia, 1977). The nature of groundwater has received a matter of huge concern due to heavy metal contamination because of ongoing industrialization and truly developing urbanization. A critical concern is tarnishing of water by trace metals. Studies have demonstrated that cardiovascular, neurological and renal problems add to substantial harmfulness of metal (Bartram and Balance, 1996). The presence of

Ghanshyam Shakar, Research Scholar, MATS University, Raipur,
e-mail: ghanshyamsakar@matsuniversity.ac.in
Bhumika Das, Associate Professor, MATS University, Raipur, Chhattishgarh, India

https://doi.org/10.1515/9783110721355-004

fluoride nitrates, arsenic, cadmium, lead and other poisonous metals are very danger-
ous to human beings (CGWB, 2010). The quick development of industrialization and urbanization has delivered an unconstructive impact on the climate over the last 20 years.
The filtering cycle has been spoiled by business, civil and agrarian waste containing
pesticides, bug sprays, manure buildups and hefty metals containing groundwater
water. These toxins are acquainted with the groundwater and soil framework through
various human exercises and fast industrialization development that straightforwardly
or by implication influence human wellbeing (Sharma Supriya et al., 2016).

4.2 Study area

Chhattisgarh is one of the 29 states of India, situated in the middle east of the nation. It is the 10th biggest state in India, with an area of 135,191 km^2 (52,198 sq mi).
With a population of 25.5 million, Chhattisgarh is the seventeenth most populated
state in the nation. An asset rich state, it is a wellspring of power and steel for the
nation, representing 15% of the absolute steel created. Chhattisgarh is one of the
quickest creating states in India. The northern and southern parts of the state are
hilly, while the middle part is a productive plain. The most elevated point in the
state is the Gaurlata. The atmosphere of Chhattisgarh is tropical. It is hot and moist
due to Tropic of Cancer and its reliance on the monsoon for precipitation. Summer
temperatures in Chhattisgarh can arrive at 45 °C (113 °F). The rainstorm season is
from late June to October. Chhattisgarh gets a normal of 1,292 mm (50.9 in) of precipitation. Winter is from November to January, and it is a decent ideal opportunity
to visit Chhattisgarh. Winters are lovely with low temperatures and less moistness.

Raipur is a city in the Raipur district of the Indian state of Chhattisgarh. It is the
capital of the Chhattisgarh. The Raipur region covers a territory of 12,461.9 sq. km. It is
situated in the mid of Chhattisgarh state. It falls in the Survey of India's topo Sheet Nos.
64G/11 and **6G/12** (1: 50,000 Scale). The area is bounded by Baloda Bazar region in the
north, Durg area in the west, Raigarh region in the east and Dhamteri region in the
south. Raipur is situated close to the central point of a huge plain; this is why it is referred as "rice bowl of India", where many assortments of rice are developed. The Mahanadi River goes towards the east of the city of Raipur, and the southern side has
thick woods. The Maikal Hills ascend on the north-west of Raipur; on the north, the
land rises and converges with the Chota Nagpur Plateau, which expands north-east
across Jharkhand state. On the south of Raipur lies the Deccan Plateau. Raipur has a
tropical wet and dry atmospheric; temperature is moderate consistently, apart
from March to June, which can be incredibly hot, up to sometimes 48 °C. The city
gets around 1,300 mm of precipitation, generally in the monsoon season from late-
June to early October. Winters last from November to January and are mellow, in
spite of the fact that low temperature can tumble to 5 °C making it sensibly cold.

Siltara is a gram panchayat situated in the Raipur region of Chhattisgarh state, India. The latitude 21.3811556 and longitude 81.6637765 are the geo coordinates of the Siltara. The closest railroad station to Siltara is Mandhar which is situated in and around 6.4 km distance. Siltara's closest air terminal is Swami Vivekananda Airport arranged at 23.1 km distance.

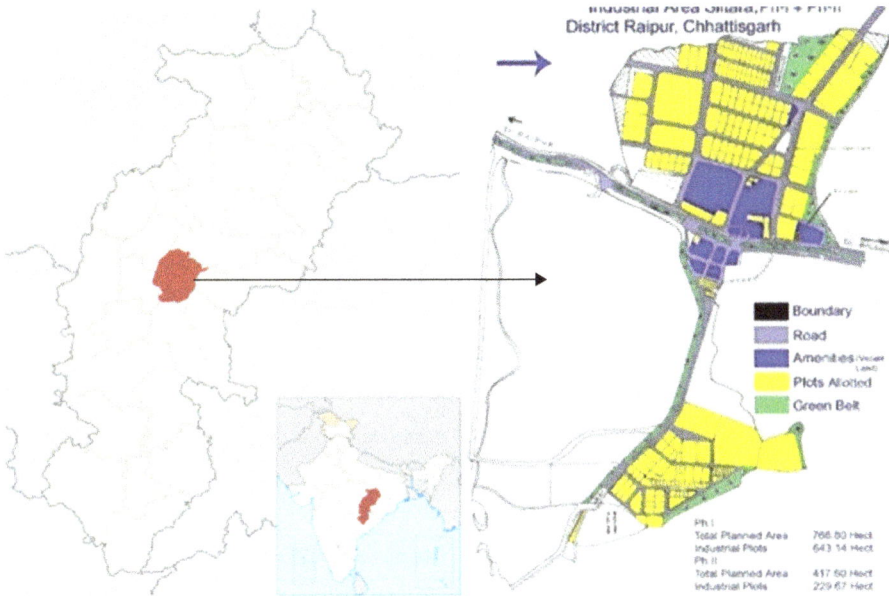

Figure 4.1: Map of study area, not to scale (source – CSIDC).

4.3 Regional geology

The Chhattisgarh Basin covers a territory of around 36,000 km^2 that covers the Bastar Craton's stone gneiss and volcanic basement (Ramakrishnan and Vaidyanadhan, 2008). The basin's southern and eastern edges have depositional cooperation with the basement rock, while the basin's western and northern edges are restricted by fault. The basin progression (~2,500 m thick) comprises essentially of subordinate conglomerate, sandstones, shale and carbonates and tuffs at various stratigraphic levels. Chattisgarh Supergroup is the succession, which is additionally separated into different groups (Murti, 1987; Das et al., 1992; Patranabis-Deb and Chaudhuri, 2008).

Nonetheless, the layers plunge tenderly at a point of 2–10° along the NS faults; the plunge goes from 20–25°. Structural disturbances along the western, northern and eastern edges are found in the Chattisgarh Basin. NNE-SSW and E-W delimit the western and northern edges of the basin to the ENE-WSW faults, individually. The deformation in the eastern part of the basin is communicated by solid open warps and enormous fault moving NNW-SSE, NNE-SSW, E-W and NE-SW inside the strata. The rejuvenated basin opening faults (Chaudhuri et al., 2002) are thought to be such huge scale fault. It was gathered that the basin shaped as an intracratonic fracture based on sedimentary gatherings, facies and stratigraphic design (Chaudhuri et al., 1999; 2002; Roy and Prasad, 2001; Patranabis-Deb and Chaudhuri, 2002 and Patranabis-Deb and Chaudhuri, 2008). Utilizing the K-Ar dating measure, Kreuzer et al. (1977) dated the authigenic glauconites from the Chaporadih Formation of the Chandarpur Group and construed the age as 700–750 Ma and viewed the progression as Neoproterozoic.

Table 4.1: Stratigraphic succession Chhattisgarh supergroup (Das et al., 1992; 2003).

Age	Supergroup	Group	Formation	Lithology
QUATERNARY	Recent to sub-recent		Alluvium and Laterite	Sand, Silt, Clay and lateritic Soil
PROTEROZOIC	Chhattisgarh Supergroup	Raipur Group	Maniyari formation	
			Hirri formation	
			Taranga formation	
			Chandi formation	Limestone, Sandstone and Shale
			Gunderdehi formation	Shale
			Charmuria formation	Limestone and Shale
		Chandrapur Group	Kanspathar formation	Sandstone, Siltstone, Shale and Conglomerate
			Choparadih formation	
			Lohardi formation	
ARCHAEAN	Basement crystallines – granite, gneisses, granulite and amphibolite			

4.3.1 Local geology

It falls under the Proterozoic Chandi formation of the Raipur Group of the Chhattisgarh Super Group, according to the nearby topography of the Siltara region. Limestone, shale, sandstone and dolerite intrusion are the primary rock types that occurs in this district in certain zones.

Table 4.2: Stratigraphic succession of the study area.

Age	Supergroup	Group	Formation	Lithology
Proterozoic	Chhattisgarh Supergroup	Raipur Group	Chandi Formation	Limestone, Sandstone & Shale

4.4 Methodology

4.4.1 Sample collection

Tests from different sources, for example, burrowed wells, borewells, lakes and drain-pipe parts (pre-storm and post-rainstorm) should be gathered to explore the effect of chemical on the groundwater. The fluid samples ought to be gathered soon after assortment in washed 1 L new polyethylene bottles and fermented (5 ml for each litre) with HNO_3. Sample for the evaluation of biochemical oxygen demand (BOD) and chemical oxygen demand (COD) should be acquired independently and not acidified. Until it is acidified, the electrical conductivity (EC) and pH of fluid examples can be estimated in the area. Five samples ought to be taken for every specimen type. Tests ought to be taken to the research centre and set in a cooler before further testing is done.

4.4.2 Sample preparation

While setting up the samples of groundwater, it must be filtered by Whatman 41 filter paper to eliminate any suspended strong particles. For heavy metal examination utilizing total reflection X-ray fluorescence (TXRF) procedure, 10 mL of each kind of test must be taken in a plastic vial and should be additionally fermented with 0.5 mL analar grade HNO_3. It must be kept under observation for 16 h for complete disintegration of inorganic salt contained in more modest strong particles that may be available in the sample even after filtration. Then 3 mL of the each sample must be internally normalized with 6 mL of standard yttrium arrangement (E.Merck, Germany).

4.4.3 Physical characteristics

Physical characteristics that need to be examined are: appearance, colour, pH, temperature, odour, electrical conductivity (EC), total suspended solids (TSS), total dissolved solids (TDS), total hardness and turbidity. With a pH/temperature meter, the pH and temperature can be assessed in situ. It is conceivable to test the colour by mixing the composite samples until the sediments are fully suspended and then be adjusted utilizing a colour disk. The Jenway M470 Portable Conductivity/TDS meter can be utilized to survey TDS, TSS, EC and turbidity in situ.

4.4.4 Chemical and organic characteristics, heavy metals

Alkalinity, acidity, nitrate, chloride, sulphate, phosphate, dissolved oxygen (DO), BOD, COD, phenol compounds, oil and grease, As, Al, Zn, Fe, BP, Cu, Ni, Mn, Cr, Cd, Mg, Ca and cyanide are included in the chemical and organic characteristics. According to American public health association (APHA), dissolved oxygen can be assessed using the Winkler method with azide modification. By subtracting the value of the final DO concentration (after 5 days of incubation at 200 °C) from the initial DO concentration, the BOD can be analysed. COD can be found out by the technique of dichromate reflux because it has an advantage over oxidants due to its oxidizing strength and its applicability to a wide range of samples. It is possible to spectrophotometrically test nitrate, phosphate and sulphate. Using titrimetric method, chloride can be assessed. It is possible to test oil and grease, phenol compounds, alkalinity, acidity and cyanide using methods adapted from traditional water and wastewater analysis methods (APHA, 1992).For the determination of heavy metals and certain trace metals, including Cd, Cr, Mn, Ni, Cu, Pb, Fe, Zn, Al, As, Ca and Mg, an atomic absorption spectrophotometer (Boston, MA 02118-2512, USA) may be used (Ogwo et al., 2014).

4.4.5 Bacteriological characteristics

Analysis of bacteriological characteristics involves total plate count, total coliform and *Escherichia coli*. It is possible to calculate the total plate count using the heterotrophic plate count introduced by the APHA. Using the most portable number (MPN) method, complete coliform and *E. coli* analysis can be done. The method followed by three successive steps: a presumptive test, a verified test and a full test that recognises coliform bacteria as a faecal contamination indicator (APHA, 1998). Results can be analysed statistically and compared with WHO (Ogwo et al., 2014).

Table 4.3: Physical and chemical properties of the tube well as per IS 10500-2012.

S. no.	Parameter	Unit	Accept. limit	Permi. limit
1	Colour	Hazen Unit	5	15
2	Odour		Agreeable	Agreeable
3	pH		6.5–8.5	No relaxation
4	Turbidity	NTU	1	5
5	Total Dissolved Solids	mg/L	500	2,000
6	Ammonia	mg/L	0.5	No relaxation
7	Boron	mg/L	0.5	1

Table 4.3 (continued)

S. no.	Parameter	Unit	Accept. limit	Permi. limit
8	Calcium	mg/L	75	200
9	Chloride	mg/L	250	1,000
10	Fluoride	mg/L	1	15
11	Magnesium	mg/L	30	100
12	Nitrate	mg/L	45	No relaxation
13	Total Alkalinity	mg/L	200	600
14	Sulphate	mg/L	200	400
15	Total Hardness	mg/L	200	600
16	Temperature	°C	–	
17	Sodium	mg/L	–	
18	Iron	mg/L	0.3	No relaxation
19	Cadmium	mg/L	0.003	No relaxation
20	Chromium	mg/L	0.05	No relaxation
21	Zinc	mg/L	5	1.5
22	Manganese	mg/L	0.1	0.3
23	Nickel	mg/L	0.02	No relaxation

4.5 Sample collection and preparation

During January–February 2020, systematic sampling was conducted. In pre-washed polythene, narrow mouth, well, borewell bottles, a total of 15 groundwater samples were gathered. Before taking samples, the bottles were rinsed twice. Special attention was given to all those areas where the contamination of fluoride was expected. Based on this report, representative wells were selected. The water samples from the borewells were obtained after pumping out water for about 10 min to extract stagnant water from the borewell.

4.6 Analysis

Hanna metres were used to calculate some significant physical parameters such as temperature, pH, reduction potential (RP), EC, TDS (Model nos. HI 8424, HI 9142, HI

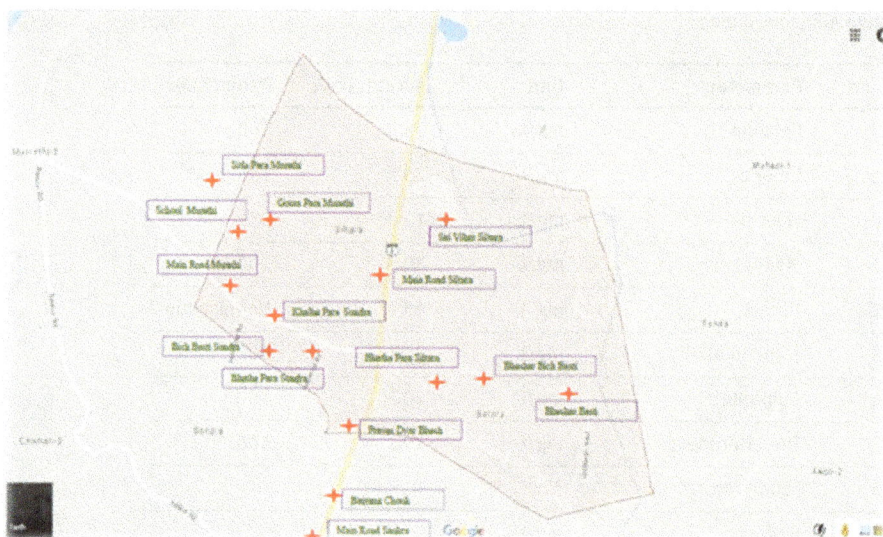

Figure 4.2: Sampling location map.

991300). Groundwater hardness measured as total hardness using ethylenediaminete-traacetic acid (EDTA; 0.01 M) complex metric titration with Eriochrome Black-T (EBT) and pH 10 buffer solution as an indicator and groundwater alkalinity measured using H_2SO_4 (0.02 N) titration with 1 mL NaOH and three drops of phenolphthalein indicator. The concentration of calcium and magnesium ions was measured using the EDTA titration method with (1 N) NaOH and the P&R indicator pinch. The fluoride ion concentration was determined using by ion-selective electrode method (Metrohm ion meter – 781) with a 1:1 total ion strength adjustment buffer (TISAB). The buffer preparation was added by 58 g NaCl + 5 g CDTA (trans-2,2-*NNNN*- cyclodiamine tetraacetic acid) + 57 mL glacial acetic acid and adjusts to near 5.5 pH with 8 M NaOH then makeup with 1 L ultra-pure distilled water. The concentration of fluoride ions was measured using the ion-selective electrode method (Metrohm ion metre-781) with a 1:1 TISAB. A total of 58 g NaCl + 5 g CDTA (trans-2,2-*NNNN*- cyclodiamine tetraacetic acid) + 57 mL glacial acetic acid and changes of near 5.5 pH with 8 M NaOH were applied to the buffer preparation and then 1 L of ultra-pure distilled water was added.

NO_3^- and NH_4^+ measurements were calculated using the ion-selective electrode (Metrohm ion metre-781) 1:1 buffer (1 mol /L $(NH_4)_2SO_4$ and NaOH) process.

The concentration of Na^+ and K^+ ions was measured by a flame photometer (SYSTRONIC Flame photo meter-130). The SO_4^{2-} ions were calculated as a pinch of $BaCl_2$ by the turbidity metre (digital turbidity metre, Model No. 331). The chloride ion was calculated by chromate indicator with Mohr solution.

NO_3^- and NH_4^+ measurements were calculated using the ion-selective electrode (Metrohm ion metre-781) 1:1 buffer (1 mol /L $(NH_4)_2 SO_4$ and NaOH) process.

The concentration of Na$^+$ and K$^+$ ions was measured by a flame photometer (SYSTRONIC Flame photo meter-130). The SO$_4^{2-}$ ions were calculated as a pinch of BaCl$_2$ by the turbidity metre (digital turbidity metre, Model No.331). The chloride ion was calculated by chromate indicator with Mohr solution.

Table 4.4: Sampling location.

S. No	Location	Age	Depth in feet
1	Pravesh Dvar Bheshar	10	110
2	Bheshar Basti	10	100
3	Bheshar Bich Basti	20	30
4	Khalhe Para Sondra	35	150
5	Bich Basti Sondra	90	40
6	Bhatha Para Sondra	17	200
7	Bajrang Chouk Sankra	15	205
8	Main Road Sankra	6	200
9	Bhatha Para Siltara	20	250
10	Sai Vihar Siltara	6	65
11	Main Road Siltara	20	215
12	Main Road Murethi	1	324
13	School Murethi	30	306
14	Goura Para Murethi	10	207
15	Sitla Para Murethi	10	297

4.7 Result

4.7.1 Physical parameter analysis

Table 4.5: Analysis of physical parameters.

S. no.	Location	T °C	pH	EC (µs)	TDS (mg/L)	RP (MV)	TH (mg/L)	Alkalinity (mg/L)	Mg$^+$ (mg/L)
1	PravesDvarBheshar	26.4	6.30	1,374	683	250.4	235	730	18.36
2	Bheshar Basti	25.9	6.24	1,308	652	237.7	275	720	17.32
3	BhesharBich Basti	25.5	6.58	1,747	872	215.3	230	850	30.58

Table 4.5 (continued)

S. no.	Location	T °C	pH	EC (µs)	TDS (mg/L)	RP (MV)	TH (mg/L)	Alkalinity (mg/L)	Mg⁺ (mg/L)
4	Khalhe Para Sondra	25.4	6.26	1,366	681	222.1	245	650	12.90
5	Bich Basti Sondra	25.5	6.24	1,439	717	245.8	250	730	17.06
6	Bhatha Para Sondra	25.3	6.25	1,041	518	253.9	255	550	10.56
7	Bajrang ChoukSankra	25.3	6.45	707	352	246.5	110	610	15.50
8	Main Road Sankra	25.3	6.01	1,222	611	285.5	165	720	22.78
9	Bhatha Para Siltara	25.3	6.05	1,208	602	272.3	225	600	14.98
10	Sai ViharSiltara	25.4	6.09	874	437	274.1	265	600	15.76
11	Main Road Siltara	25.5	6.42	1,021	509	259.9	170	650	17.32
12	Main Road Murethi	25.4	6.22	4.53	2.26	274.5	180	750	12.38
13	School Murethi	25.7	6.13	932	463	274.5	230	850	26.94
14	Goura Para Murethi	25.5	6.34	1,011	509	262.4	200	680	17.84
15	Sitala Para Murethi	26.0	6.13	1,667	831	264.3	345	850	17.84

4.7.2 Chemical parameters analysis

Table 4.6: Analysis of chemical parameters.

	Location	No_3^-	PO_4^{3-}	F^-	Cl^-	Na^+	Ca^{2+}	K^+	SO_4^{2-}	NH_4^+
1	PravesDvarBheshar	6.485	0.510	4.90	896	75.61	16.11	1.59	60.15	16.2
2	BhesharBasti	9.023	0.567	4.66	854	71.33	15.10	1.59	54.37	16.2
3	BhesharBichBasti	19.95	0.870	7.40	874	125.9	17.62	12.46	100.1	14.3
4	Khalhe Para Sondra	20.08	0.567	3.80	836	53.12	15.10	41.91	100.5	17.4
5	BichBasti Sondra	19.98	0.784	3.26	754	70.26	17.62	47.94	101.8	15.1
6	Bhatha Para Sondra	13.95	0.822	3.20	658	43.49	13.08	0.25	123.6	15.8
7	BajrangChoukSankra	13.53	0.728	4.64	516	63.83	13.59	5.76	39.94	11.9
8	Main Road Sankra	20.18	0.605	3.12	680	93.81	15.10	17.81	60.56	17.2
9	Bhatha Para Siltara	19.58	0.558	2.72	738	61.69	15.60	1.08	59.32	12.6
10	Sai ViharSiltara	20.01	0.605	3.24	726	64.90	17.11	2.41	122.0	15.5

Table 4.6 (continued)

	Location	No$_3^-$	PO$_4^{3-}$	F$^-$	Cl$^-$	Na$^+$	Ca^{2+}	K$^+$	SO$_4^{2-}$	NH$_4^+$
11	Main Road Siltara	4.416	0.605	3.12	658	71.33	10.06	0.25	92.32	11.8
12	Main Road Murethi	8.813	0.567	4.06	472	50.98	16.11	2.26	91.08	8.16
13	School Murethi	19.57	0.614	4.18	668	110.9	15.10	1.59	58.09	10.1
14	Goura Para Murethi	18.95	0.662	4.12	762	73.47	14.09	0.41	93.15	13.4
15	Sitala Para Murethi	19.93	0.614	4.80	1,018	73.47	24.66	64.67	171.9	11.1

4.7.3 WHO prescriptions against observed value of physical and chemical species

Table 4.7: Comparison between standard and observed values.

S. no.	Species	WHO prescriptions, mg/L	Observed value (low to high value mg/L)	Average value
1	Temperature	7 °C–11 °C	25 °C–26.4 °C	25.6033
2	pH	6.5–8.5	6.01–6.68	6.275
3	EC	500	4.53–1,973	1,056.551
4	TDS	600	2.26–985	527.508
5	RP	–	208.7–285.5	251.87
6	TH	100–500	110–450	227.5
7	Mg$^+$	30	2.5–30.58	14.68
8	Alkalinity	300	490–850	653.33
9	NO$_3^-$	45	4.416–20.45	16.98
10	PO$_4^{3-}$	5	0.51–0.87	0.616
11	F$^-$	1.5	2.58–7.4	3.528
12	Cl$^-$	250	280–1,404	749.4
13	Na$^+$	20	10.3–125.94	60.45
14	Ca^{2+}	75	9.57–36.75	18.04
15	K$^+$	25	0.26–64.68	7.60
16	SO$_4^{2-}$	200	5.70–171.93	73.33
17	NH$_4^+$	–	8.16–22.30	13.30

4.7.4 Correlation matrix of ions for Siltara

Table 4.8: Correlation matrix.

	T°C	pH	EC (µs)	TDS (mg/l)	RP (MV)	TH	Alkalinity	No_3^-	PO_4^{3-}	F^-	Cl^-	Na^+	Mg^{2+}	Ca^+	K^+	SO_4^{2+}	NH_4^+
T°C	1																
pH	0.34	1															
EC (µs)	0.23	−0.10	1														
TDS (mg/l)	0.23	−0.10	1.00	1													
RP(MV)	−0.21	−0.43	−0.43	−0.43	1												
TH	0.29	−0.14	0.72	0.72	−0.35	1											
Alkalinity	0.20	0.00	0.11	0.10	0.16	−0.15	1										
No_3^-	−0.22	−0.26	0.28	0.28	−0.03	0.19	−0.03	1									
PO_4^{3-}	−0.28	0.33	0.18	0.18	−0.19	−0.07	0.23	0.11	1								
F^-	0.14	0.27	0.23	0.22	−0.18	0.02	0.65	−0.20	0.46	1							
Cl^-	0.29	−0.08	0.88	0.88	−0.45	0.83	−0.10	0.20	−0.06	0.04	1						
Na^+	0.04	−0.11	0.59	0.59	−0.11	0.21	0.61	0.13	0.34	0.55	0.41	1					

Mg²⁺	0.04	−0.11	0.59	−0.11	0.21	**0.61**	0.13	0.34	0.55	0.41	1	1	1			
Ca⁺	0.18	−0.12	**0.61**	−0.33	0.83	−0.24	0.36	−0.18	−0.14	**0.80**	0.13	0.13	1			
K⁺	0.04	−0.17	0.39	−0.04	0.22	0.44	0.25	0.22	0.26	0.18	0.20	0.20	0.09	1.00		
SO₄²⁺	0.10	−0.08	0.51	−0.12	0.47	0.31	0.16	0.33	0.20	0.47	0.44	0.44	0.26	0.52	1.00	
NH₄⁺	0.05	−0.17	0.43	−0.35	0.32	−0.11	0.17	0.05	0.01	0.25	0.24	0.24	0.15	0.11	0.28	1

Different physical parameter of groundwater(Siltara)

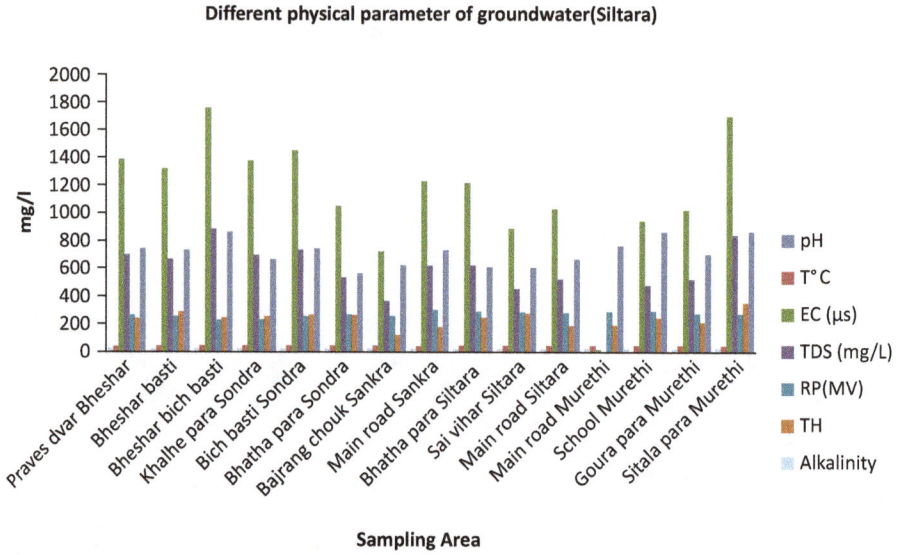

Figure 4.3: Physicochemical parameters (pH, Tem, EC,TDS, RP, T-H, H,T-A) of groundwater graph around the Siltara area.

Different chemical parameter of groundwater(Siltara)

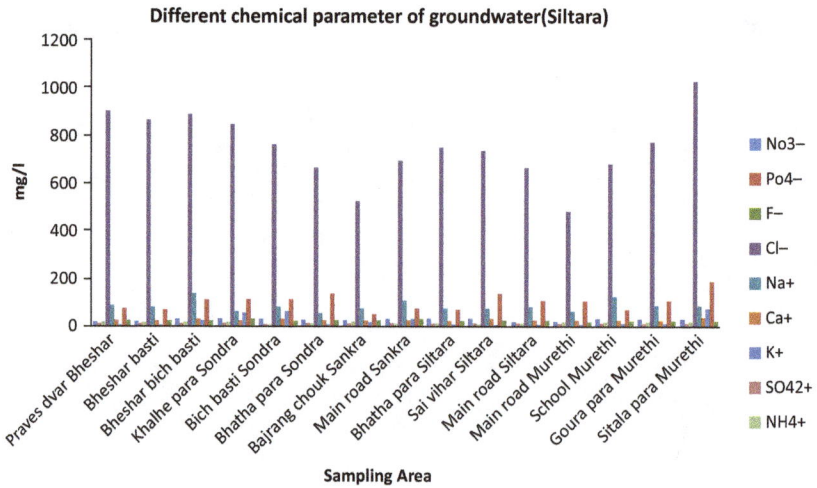

Figure 4.4: Physicochemical parameters (NO_3^-, PO_4^{3-}, F^-, Cl^-, Na^+, Ca^{2+}, K^+, SO_4^{2-}, NH_4^+) of groundwater graph around the Siltara area.

4.8 Conclusion

The hydrochemical analysis shows that except for the few areas, the current status of groundwater in Siltara is not appropriate for drinking purposes but could deteriorate in the future. It is evident from the very high percentage of water tested, the desirable limits according to WHO standards and almost approaches the maximum permissible level. Adequate effluent treatment techniques and appropriate methods of disposal are required. Both areas demonstrate significant leaching into the groundwater of various chemical components leading to the enrichment of numerous anions and cations, which ultimately suggests contamination from foreign sources. Therefore, it is time for the new definition of "water supply management" to shift to "water demand management". The key driver for improved groundwater quality in India should be effective management of effluent treatment and solid waste management, rather than furnishing various subsidies.

References

APHA, AWWA. WEF Standard Methods for the Examination of Water and Waste Water, 21st ed, American Public Health Association, Washington DC.

American Public Health Association (APHA) (1992) Standard Methods for Examination of Water and Wastewater. 18th Edition, APHA, AWWA, WPCF, NY, Washington DC.

Chaudhuri, A.K., Mukhopadhyay, J., Patranabis-Deb, S. And chanda, S.K. (1999): the Neoproterozoic Cratonic Successions Of Peninsular India. Gondwana res., v.2, pp.213–225.

Chaudhuri, A.K., Saha, D., Deb, G.K., Patranabis-Deb, S., Mukherjee, M.K. and Ghosh, G. (2002): the Purana Basins Of Southern Cratonic Province Of India: A Case For Mesoproterozoic Fossil rifts. Gondwana Res., v.5, pp.23–33.

Das, D.P., dutta, N.K.N., dutta, D.R., thanavelu, C. And baburao, K. (2003): singhora group-the oldest proterozoic lithopackage of eastern bastar craton and its significance. Indian minerals, v.57, pp.127–138.

Bartram, J., Balance, R. (1996). Water quality monitoring – A practical guide to the design and implementation of freshwater quality studies and monitoring programmes. Published on behalf of United Nations Environment Programme and the World Health Organization.

Board of Indian Standards (BIS) (2012). Indian standards for drinking water specification, (BIS10500:2012).

Brochure, C.G.W.B., (2010). Jaipur District.

Ogwo, P.A., et al. (2014). Impact of industrial effluents discharge on the quality of Nwiyi River Enugu South Eastern Nigeria. IOSR Journal of Environmental Science. Toxicology and Food Technology, (IOSR-JESTFT) e-ISSN: 2319–2402,p- ISSN: 2319–2399, 8(11), Ver. I (Nov. 2014), 22–27.

Y, C., Rajwade, R.P. (2015). Analysis of water quality in different locations of BALCO industrial area of Korba, Chhatisttgarh, India, Inter. Journal of Research in Engineering, Science and Technologies, 1, 45–49.

Punmia, B.C., (2018). Water Supply Engineering, Standard book house, New Delhi, India.
Supriya, S., et al. (2016 June). Impact of industrial development on groundwater quality- a case study of impact of effluent from Vishwakarma industrial area, Jaipur on ground water. Bulletin of Environment, Pharmacology and Life Sciences, 5(7), 12–15.
Ramakrishnan M and Vaidyanadhan R 2008 Geology of India, Bangalore; Geol. Soc. India 1&2 552.

Pallavi Pradeep Khobragade, Ajay Vikram Ahirwar

5 Seasonal variation of PM$_{10}$ in the ambient air over an urban industrial area

Abstract: In this study, PM$_{10}$ samples were collected in Raipur, India, from January 2016 to November 2016 by using respirable dust sampler. The annual average concentrations and associated standard deviation of PM$_{10}$ was 182.69 ± 90.10 µg/m^3 which was crossing the limits set by NAAQS, India, of 60 µg/m^3 for PM$_{10.}$ It was found that the annual average PM$_{10}$ concentration was higher during winter (225.13 ± 104.75) as compared to summer (168.82 ± 27.51) and monsoon (103.88 ± 29.32). Temperature inversion and low mixing heights leads to an increment in concentration while the precipitation during monsoon reduces the concentration due to wet deposition. The interrelationship between meteorological parameters and PM$_{10}$ concentration was performed and a negative correlation was found between PM$_{10}$ and wind speed (r = −0.48), temperature (r = −0.38), rainfall (r = −0.30) and humidity (r = −0.26). The average temperature, wind speed and humidity during summer (34.14 °C; 11.64 km/h; 25.14%, repectively), monsoon (27.64 °C; 11.91 km/h; 79%, respectively) and winter (22.6; 6.76 km/h; 52.76%, respectively) agree the negative correlation between the seasonal concentrations of PM$_{10}$ and relevant meteorological parameters. Significant seasonal differences in particulate matter (PM) concentrations have been observed in the course of study period. The major local festivals in the region contributing to increase in PM concentration accompany the sudden increase of particulate concentration along with favourable meteorological conditions for increase during winter. Attribution of particulates in the rapidly growing Indian cities results from industrialization, vehicular pollution, biomass burning and resuspension of road dust.

5.1 Introduction

Air pollution has been observed as the 5th leading risk factor for mortality as per the states of global air report (SOGA, 2019). Developing countries are experiencing deteriorated urban air quality, due to urbanization, industrialization (Gupta et al., 2007) biomass and fossil fuel burning, changes in land use (Nagar et al., 2019), having severe impacts on public health (Deshmukh et al., 2013; Khillare and Sarkar, 2012) causing premature deaths (Kaushik et al., 2019). Particulate matter (PM) from urban locations depends upon local and distant sources, meteorological parameters and industrial

Pallavi Pradeep Khobragade, Ajay Vikram Ahirwar, Department of Civil Engineering, National Institute of Technology, Raipur 492001, Chhattisgarh, India, e-mail: khobragadepallavi19@gmail.com

https://doi.org/10.1515/9783110721355-005

pollutants (Mukherjee and Agrawal, 2018). The increasing air pollution leads to adverse impacts on respiratory and cardiovascular systems (Deshmukh et al., 2013). Particulate matters less than 10 μm (PM_{10}) have an ability to enter the human body through respiration (Das et al., 2015) after absorbance of toxic substances emitted due to anthropogenic activities causing severe respiratory and cardiovascular diseases (Gao and Ji, 2018) and are considered as local and urban scale problems. The ambient outdoor air pollution found to cause 2.4 million deaths in 2016 (WHO, 2018) being one of the largest environmental threat.

The earth's energy budget is fluctuated by particulate matter (PM) due to scattering and absorbing radiation modifying the earth's temperature and microphysical properties of clouds (Talbi et al., 2018). Epidemiological studies have shown the increase in lung cancer, morbidity and cardiopulmonary diseases (Deshmukh et al., 2013), premature and inopportune death (Jaiswal et al., 2019) due to exposure with PM_{10}. Influencing factors in PM_{10} concentrations include high urban traffic volume, construction and demolition activities, soil dust, vehicular and industrial activities (Ganguly et al., 2019). The meteorological conditions like wind speed and wind direction, temperature, rainfall and humidity also influence the particulate concentration (Ahirwar and Bajpai, n.d.; Begum et al., 2008). The Environmental Protection Agency suggests importance of PM_{10} aerosols in determination of air quality index (Deshmukh et al., 2012).

These studies have highlighted the importance of particulate concentrations, its distribution and impacts in the developing countries like India. In the current study, particulate monitoring of PM_{10} was carried out during 2016 and compared with the National Ambient Air Quality Standards (NAAQS) of India. Variations in seasonal patterns of PM_{10} in urban weather were determined by examining the role of meteorological parameters.

5.2 Materials and methods

5.2.1 Study area

The capital of state Chhattisgarh, Raipur (21°14′ N and 81°38′ E) was selected for PM_{10} sampling located at 303.36 m AMSL. Raipur is the capital of state surrounded by industries on its eastern and western directions at about 25 km periphery. Mixed pollutants are detected due to presence of Bhilai steel plant, cement factories, iron and steel plants and many fertilizer plants surrounding the Raipur city producing particulate matter affecting surrounding environment. A number of rice mills are running in the neighbouring cities such as Durg, Korba, Dhamtari including others. Korba is an industrial city located at around 160 km from Raipur operating coal mines and thermal power plants. Urbanization and industrialization lead to high level of particulate matter concentrations in the area.

5.2.2 Particulate monitoring

Gravimetric analysis was used to determine particulate concentrations. Continuous measurement of PM$_{10}$ was carried out by using a respirable dust sampler, Envirotech instrument Pvt. Ltd., New Delhi, India using 8″ × 10″ Whatman glass fibre filters. The filter papers were kept in dessicators for around 24 h before and after sampling to remove moisture content and weighted using an analytical balance (Model: Citizon CX220). Care was taken while moving the conditioned papers from laboratory to the sampling site. The sampler was run for 24 h and PM$_{10}$ concentrations were calculated gravimetrically. The filter papers were wrapped in aluminium foil in envelopes and brought to laboratory for further analysis.

5.3 Results and discussions

5.3.1 PM$_{10}$ variation

The maximum PM$_{10}$ concentration during the study was perceived on 28 November 2016 (505.00 µg/m^3) having minimum temperature (21 °C) at a very low wind speed of 4 km/h showing humidity of about 41%. This decrease in temperature during the winter season enables the people to elevate the temperature through wood burning, biomass burning and cow dung cake as bonfire ultimately leading to elevation in pollutant concentrations (Deshmukh et al., 2013). Data collection was done to record temporal changes on PM$_{10}$ concentrations and their variation with meteorological parameters and anthropogenic activities. The annual average PM$_{10}$ concentration was (182.68 ± 90.10 µg/m^3) exceeding thrice the annual standard specified by Indian NAAQS. Local anthropogenic activities produces majority of pollutants at low altitudes (Ganguly et al., 2019). Another peak is observed on 10 November (480.17 µg/m^3) after one of the major Indian festival Diwali followed by 30 October (390.00 µg/m^3) on the day of festival.

Figure 5.1 indicates the relationship between PM$_{10}$ concentrations, wind direction and wind speed indicating that wind direction and wind speed played an important role in dispersion and dilution of pollutants. It can be seen that with higher wind speeds during the months from March to May, PM$_{10}$ concentrations were comparatively lower than that during low wind speeds from October to February when maximum concentrations were detected. When the wind flowed from NE during winter, higher concentrations were observed while the direction changed to SW and NW in summer fewer concentrations were reported. The seasonal average, maximum and minimum PM10 concentrations are given in Table 5.1. The maximum concentrations were observed during winter season and minimum during monsoon period.

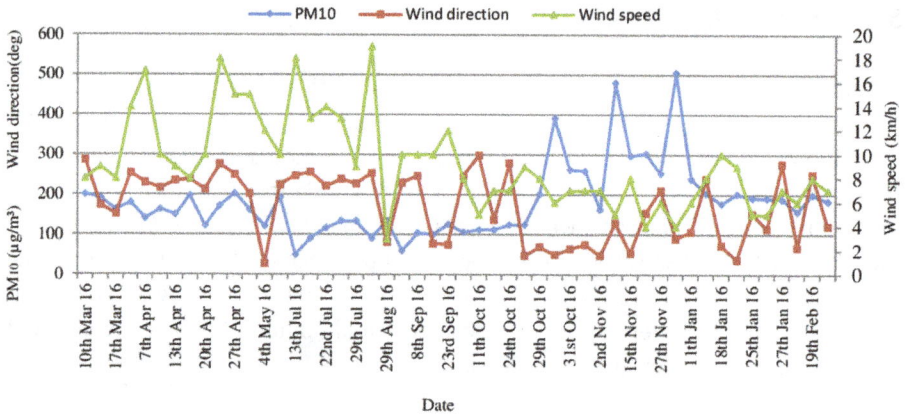

Figure 5.1: PM$_{10}$ variation (µg/m^3) with wind direction and wind speed.

Table 5.1: Seasonal average, maximum and minimum concentrations of PM10 (µg/m^3).

Season	Summer	Monsoon	Winter
Mean ± SD	168.82 ± 27.51	103.88 ± 29.32	225.13 ± 104.75
Maximum	202.90	134.41	505.00
Minimum	122.03	49.53	106.75

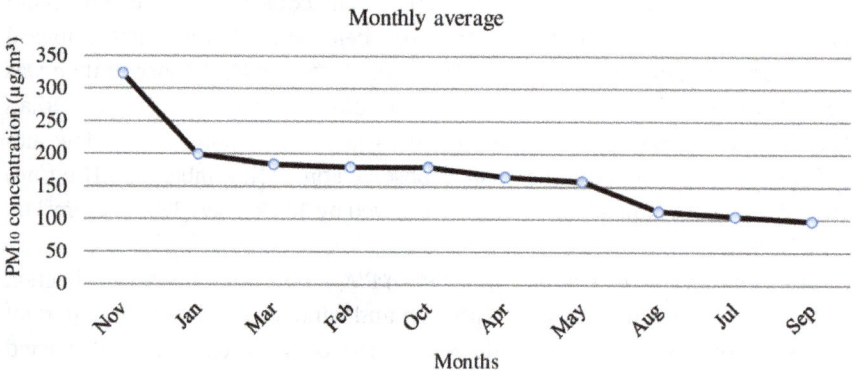

Figure 5.2: Monthly average PM$_{10}$ concentrations (µg/m^3).

The monthly averages of PM$_{10}$ range from 97.83 µg/m^3 to 322.92 µg/m^3 in the area shown in Figure 5.2. The monthly mean mass concentrations of inhalable PM$_{10}$ particles were found to be highest during winter in the month of November (322.92 µg/m^3) followed by January (198.72 µg/m^3) and then by March (183.63 µg/m^3) with minimum concentration during monsoon in the month of September (97.83 µg/m^3). Similar seasonal variations were observed and reported in other studies as (Chaudhari et al., 2012; Deshmukh et al., 2013; Gao and Ji, 2018; Nagar et al., 2019; Tiwari et al., 2015). The PM$_{10}$ concentrations were compared with other studies in India and abroad and are summarized in Table 5.2.

Table 5.2: Summary of PM$_{10}$ concentrations in India and abroad.

Location	Study period	PM$_{10}$ (µg/m^3)	Reference
Raipur (India)	June 2009–Juyl 2010	109.8– 455.6	Deshmukh et al. (2013)
Lucknow (India)	Summer	107.6–237.8	Sharma et al. (2006)
Nagpur (India)	2006	100–254	Chaudhari et al. (2012)
Delhi (India)		268.6	Shandilya et al. (2007)
Mumbai (India)	April 2003–March 2004	61	Kumar and Joseph (2006)
Calicut (India)	February 2017–March 2017	29.17–129.17	Keerthi et al. (2018), p. 200
Trivandrum (India)	October 1998–December 2000	49.57	Pillai et al. (2002)
Peshawar (Pakistan)	40,634	Max: 553 ± 101; Min: 410 ± 95	Alam et al. (2015)
Korea	2013	45.0 ± 20.4	Vellingiri et al. (2015)
Hong Kong (China)	October 2004 to September 2005	81.3 ± 37.7	Gao and Ji (2018)

5.3.2 Back trajectory analysis

To examine the long range transport of PM$_{10}$ concentrations, back trajectory analysis was performed for five consecutive days during maximum concentration days at 500 m, 1,000 m and 2,000 m using hybrid single particle lagrangian integrated trajectory (HYSPLIT) indicating trajectory routes (Figure 5.3.) using the meteorological data of national oceanic and atmospheric administration (NOAA; website: https://www.ready.noaa.gov/hypub-bin/trajresults.pl?jobidno=133311). Using this model, 120 h of backward trajectories were calculated ending at 12:00 UTC. The analysis implies the prevailing wind directions on the maximum concentrations days from south and south-west. At low altitudes, the concentration is due to local anthropogenic activities during Diwali festival in which huge numbers of firecrackers are burnt. This is accompanied with local wood and biomass combustion to elevate the temperature during winter time. Similar studies in Delhi reported higher PM$_{10}$ concentrations during local festivals (Perrino et al., 2011).

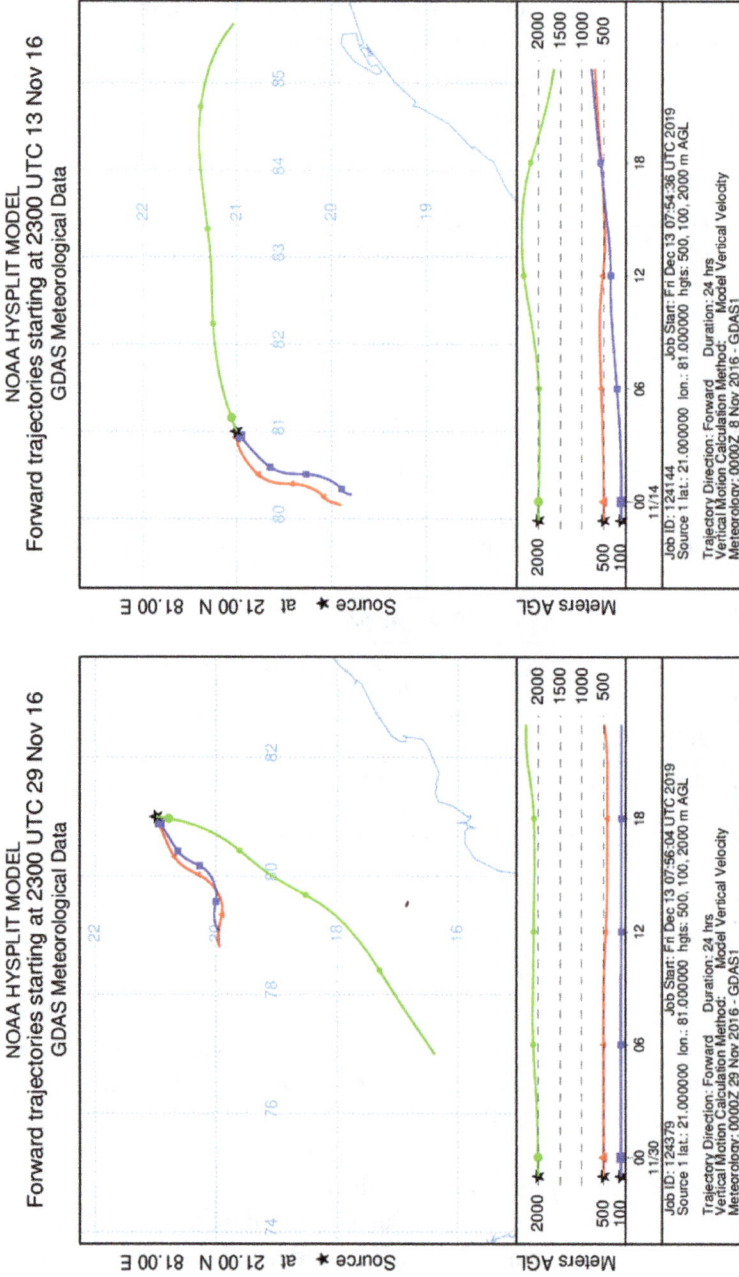

Figure 5.3: Back trajectory analysis using HYSPLIT model.

5.3.3 Meteorological parameters

The PM$_{10}$ concentrations were higher during winter season (225.13 ± 104.75 µg/m^3) followed by summer (168.82 ± 27.51 µg/m^3) monsoon (103.88 ± 29.32 µg/m^3). The higher concentrations during winter season are due to meteorological impacts governing the dispersion of pollutants in the lower atmosphere (Tiwari et al., 2015). Lower mixing heights due to inversion and low temperature with low wind speeds during winter are the foremost causes of higher concentrations due to which poor dispersion occurs trapping the particulates near the lower atmosphere (Khillare and Sarkar, 2012; Nagar et al., 2019; Tiwari et al., 2015). Wind rose diagrams during the three seasons indicating the prevailing wind directions and wind speeds are shown in Figure 5.4.

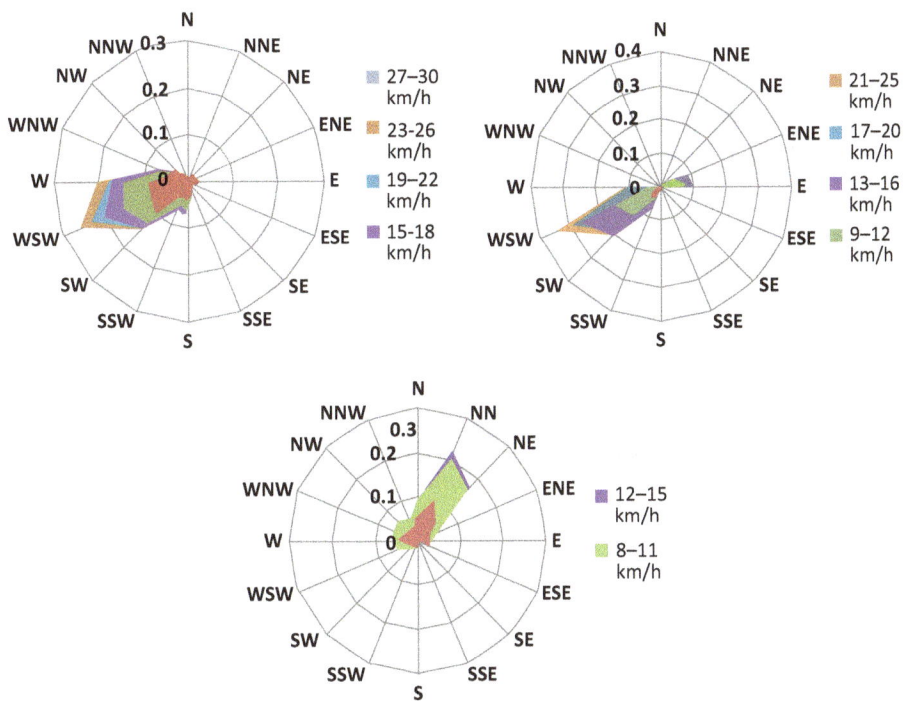

Figure 5.4: Wind rose diagram during (a) summer (b) monsoon and (c) winter.

Correlation analysis between the PM$_{10}$ concentrations and meteorological conditions was performed as shown in Table 5.3, and it was observed that wind speed is negatively correlated with PM$_{10}$ ($r = -0.48$) indicating decrease in concentration during summer with higher speed and increase in winter time during lower speed. Temperature ($r = -0.38$), rainfall ($r = -0.30$) and humidity ($r = -0.26$) also showed negative correlation with PM$_{10}$ concentrations indicating low concentrations with rainfall

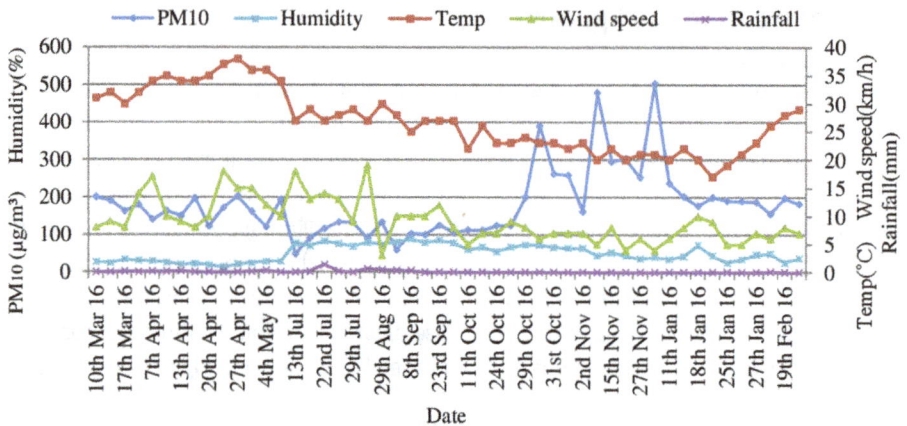

Figure 5.5: PM$_{10}$ variation with meteorological factors.

Table 5.3: Correlation analysis between PM$_{10}$ and meteorological parameters.

Variables	PM10	Temp	Wind speed	Rainfall	Humidity
PM10	1.00				
Temp	−0.38	1.00			
Wind speed	−0.48	0.57	1.00		
Rainfall	−0.30	0.16	0.34	1.00	
Humidity	−0.26	−0.42	0.02	0.32	1.00

due to settlement of particles with moisture (Seinfeld and Pandis, 2006), higher temperature and humidity (Kulshrestha et al., 2004). Figure 5.5 shows the variation of PM10 particles with changes in meteorological conditions.

5.4 Conclusion

Air quality study in the Raipur city implied that it is much inferior in comparison with similar studies in India and abroad due to urbanization and industrialization. The annual average PM$_{10}$ concentration showed that it is thrice the standards set by NAAQS established by United States Environmental Protection Agency (USEPA). It has been observed that PM$_{10}$ concentration was higher during winter (225.13 ± 104.75) in comparison to summer (168.82 ± 27.51) and monsoon (103.88 ± 29.32) due to lower boundary levels due to inversion. In summer, pollutant dispersion occurs due to higher wind speed resulting in low PM_{10} concentration while in monsoon due to wet scavenging the decrease in concentration has been noticed. Also, the increased use of wood and biomass burning leads to an increase in pollutant concentration during winter. PM_{10} showed negative correlations with temperature (–0.38), wind speed (–0.48), rainfall (–0.30) and relative humidity (–0.26) agreeing over the seasonal variations of PM. Back trajectory analysis using HYSPLIT model showed the contributions from local anthropogenic sources during the period of maximum concentration. Uncontrolled biomass burning should be stopped during winter to help maintain the PM_{10} levels in the area. Proper actions should be taken by the Chhattisgarh government to mitigate the high levels of PM_{10}.

References

Ahirwar, A.V., Bajpai, S., n.d. Seasonal Variability Of Tspm, Pm10 And Pm2.5 In Ambient Air At An Urban Industrial Area In Eastern Central Part Of India 10.

Alam, K., Rahman, N., Khan, H.U., Haq, B.S., Rahman, S. (2015). Particulate matter and its source apportionment in Peshawar, Northern Pakistan. Aerosol and Air Quality Research, 15, 634–647. Doi: https://doi.org/10.4209/aaqr.2014.10.0250.

Begum, B.A., Biswas, S.K., Hopke, P.K. (2008). Assessment of trends and present ambient concentrations of PM2.2 and PM10 in Dhaka, Bangladesh. Air Quality, Atmosphere & Health, 1, 125–133. Doi: https://doi.org/10.1007/s11869-008-0018-7.

Chaudhari, P.R., Gupta, R., Gajghate, D.G., Wate, S.R. (2012). Heavy metal pollution of ambient air in Nagpur City. Environmental Monitoring and Assessment, 184, 2487–2496. Doi: https://doi.org/10.1007/s10661-011-2133-4.

Das, R., Khezri, B., Srivastava, B., Datta, S., Sikdar, P.K., Webster, R.D., Wang, X. (2015). Trace element composition of PM2.5 and PM10 from Kolkata – a heavily polluted Indian metropolis. Atmospheric Pollution Research, 6, 742–750. Doi: https://doi.org/10.5094/APR.2015.083.

Deshmukh, D.K., Deb, M.K., Mkoma, S.L. (2013). Size distribution and seasonal variation of size-segregated particulate matter in the ambient air of Raipur city, India. Air Quality, Atmosphere & Health, 6, 259–276. Doi: https://doi.org/10.1007/s11869-011-0169-9.

Deshmukh, D.K., Deb, M.K., Verma, D., Verma, S.K., Nirmalkar, J. (2012). Aerosol size distribution and seasonal variation in an urban area of an Industrial City in central India. The Bulletin of Environmental Contamination and Toxicology, 89, 1098–1104. Doi: https://doi.org/10.1007/s00128-012-0834-1.

Ganguly, R., Sharma, D., Kumar, P. (2019). Trend analysis of observational PM10 concentrations in Shimla city, India. Sustainable Cities and Society, 51, 101719. Doi: https://doi.org/10.1016/j.scs.2019.101719.

Gao, Y., Ji, H. (2018). Microscopic morphology and seasonal variation of health effect arising from heavy metals in PM2.5 and PM10: One-year measurement in a densely populated area of urban Beijing. Atmospheric Research, 212, 213–226. Doi: https://doi.org/10.1016/j.atmosres.2018.04.027.

Gupta, A.K., Karar, K., Srivastava, A. (2007). Chemical mass balance source apportionment of PM10 and TSP in residential and industrial sites of an urban region of Kolkata, India. Journal of Hazardous Materials, 142, 279–287. Doi: https://doi.org/10.1016/j.jhazmat.2006.08.013.

Health Effects Institute, (2019). State of Global Air 2019. Special Report, Boston, MA: Health Effects Institute. ISSN 2578–6873.

Jaiswal, N.K., Ramteke, S., Patel, K.S., Saathoff, H., Nava, S., Lucarelli, F., Yubero, E., Viana, M. (2019). Winter particulate pollution over Raipur, India. The Journal of Hazardous, Toxic, and Radioactive Waste, 23, 05019001. Doi: https://doi.org/10.1061/(ASCE)HZ.2153-5515.0000444.

Kaushik, G., Chel, A., Patil, S., Chaturvedi, S. (2019). Status of Particulate Matter Pollution in India. In Review, A., Hussain, C.M. (Ed.), Handbook of Environmental Materials Management, Springer International Publishing, Cham, 167–193. Doi: https://doi.org/10.1007/978-3-319-73645-7_78.

Keerthi, R., Selvaraju, N., Alen Varghese, L., Anu, N. (2018). Source apportionment studies for particulates (PM $_{10}$) in Kozhikode, South Western India using a combined receptor model. Chemistry and Ecology, 34, 797–817. Doi: https://doi.org/10.1080/02757540.2018.1508460.

Khillare, P.S., Sarkar, S. (2012). Airborne inhalable metals in residential areas of Delhi, India: distribution, source apportionment and health risks. Atmospheric Pollution Research, 3, 46–54. Doi: https://doi.org/10.5094/APR.2012.004.

Kulshrestha, U.C., Nageswara Rao, T., Azhaguvel, S., Kulshrestha, M.J. (2004). Emissions and accumulation of metals in the atmosphere due to crackers and sparkles during Diwali festival in India. Atmospheric Environment, 38, 4421–4425. Doi: https://doi.org/10.1016/j.atmosenv.2004.05.044.

Kumar, R., Joseph, A.E. (2006). Air Pollution Concentrations of PM2.5, PM10 and NO2 at Ambient and Kerbsite and Their Correlation in Metro City – Mumbai. Environmental Monitoring and Assessment, 119, 191–199. Doi: https://doi.org/10.1007/s10661-005-9022-7.

Mukherjee, A., Agrawal, M. (2018). Air pollutant levels are 12 times higher than guidelines in Varanasi, India Sources and transfer. Environmental Chemistry Letters, 16, 1009–1016. Doi: https://doi.org/10.1007/s10311-018-0706-y.

Nagar, P.K., Sharma, M., Das, D. (2019). A new method for trend analyses in PM10 and impact of crop residue burning in Delhi, Kanpur and Jaipur, India. Urban Climate, 27, 193–203. Doi: https://doi.org/10.1016/j.uclim.2018.12.003.

Perrino, C., Tiwari, S., Catrambone, M., Dalla Torre, S., Rantica, E., & Canepari, S. (2011). Chemical characterization of atmospheric PM in Delhi, India, during different periods of the year including Diwali festival. Atmospheric Pollution Research, 2(4), 418–427. Doi: https://doi.org/10.5094/APR.2011.048

Pillai, P.S., Suresh Babu, S., Krishna Moorthy, K. (2002). A study of PM, PM10 and PM2.5 concentration at a tropical coastal station. Atmospheric Research, 61, 149–167. Doi: https://doi.org/10.1016/S0169-8095(01)00136-3.

Seinfeld, J. H., & Pandis, S. N. (2006). Atmospheric chemistry and physics: From air pollution to climate change. Hoboken, New Jersey.

Shandilya, K.K., Khare, M., Gupta, A.B. (2007). Suspended Particulate Matter Distribution in Rural-Industrial Satna and in Urban-Industrial South Delhi. Environmental Monitoring and Assessment, 128, 431–445. Doi: https://doi.org/10.1007/s10661-006-9337-z.

Sharma, K., Singh, R., Barman, S.C., Mishra, D., Kumar, R., Negi, M.P.S., Mandal, S.K., Kisku, G.C., Khan, A.H., Kidwai, M.M., Bhargava, S.K. (2006). Comparison of trace metals concentration in PM10 of different locations of Lucknow City, India. The Bulletin of Environmental Contamination and Toxicology, 77, 419–426. Doi: https://doi.org/10.1007/s00128-006-1082-z.

Talbi, A., Kerchich, Y., Kerbachi, R., Boughedaoui, M. (2018). Assessment of annual air pollution levels with PM1, PM2.5, PM10 and associated heavy metals in Algiers, Algeria. Environmental Pollution, 232, 252–263. Doi: https://doi.org/10.1016/j.envpol.2017.09.041.

Tiwari, S., Pipal, A.S., Hopke, P.K., Bisht, D.S., Srivastava, A.K., Tiwari, S., Saxena, P.N., Khan, A.H., Pervez, S. (2015). Study of the carbonaceous aerosol and morphological analysis of fine particles along with their mixing state in Delhi, India: a case study. Environmental Science and Pollution Research, 22, 10744–10757. Doi: https://doi.org/10.1007/s11356-015-4272-6.

Vellingiri, K., Kim, K.-H., Ma, C.-J., Kang, C.-H., Lee, J.-H., Kim, I.-S., Brown, R.J.C. (2015). Ambient particulate matter in a central urban area of Seoul, Korea. Chemosphere, 119, 812–819. Doi: https://doi.org/10.1016/j.chemosphere.2014.08.049.

World health statistics, (2018). Monitoring health for the SDGs, sustainable development goals. Geneva: World Health Organization; 2018. Licence: CC BY-NC-SA 3.0 IGO.

Sahajpreet Kaur Garewal, Avinash D. Vasudeo

6 Groundwater sustainability assessment using multi-criteria analysis

Abstract: Groundwater being a precious natural resource is increasingly endangered, due to natural and anthropogenic activities. Considering the invisibility and complex nature of the groundwater it is difficult to manage the valuable resource. Sustainability assessment is considered as a useful technique, which provides enough information to assist management. Evaluating vulnerability of groundwater to contamination is essential to sustain the integrity of groundwater. In the present study, vulnerability of groundwater was assessed by Multi-Criteria Analysis (MCA) and compare with result of DRASTIC. The MCA is an efficient method for groundwater vulnerability assessment including groundwater quality parameters. DRASTIC is a traditional approach to identify the groundwater vulnerability considering hydrogeological characteristics of the area. The combine application of MCA and DRASTIC method was found to be effective for making efficient measure for groundwater protection.

6.1 Introduction

Groundwater is a fairly ubiquitous natural resource. Increasing urbanization, usage of chemical in daily life and careless disposal of hazardous waste are threatening the precious natural resources. It is a matter of deep concern globally, due to its harmful effect to humans and surrounding environment. Once the contamination occurs to groundwater then it is very difficult or some time impossible to take corrective steps for remediation, considering the cost and its invisibility in nature (Johnson, 1979). Proper management of groundwater is a primary concern, and if done meticulously would lead to a cleaner and healthier environment. Protection of existing surface and groundwater resources against contamination and overexploitation is one of the most important aspects of Integrated Water Resource Management (Hamutoko et al., 2016).

Numerous researches in the last few decades have been extensively carried out to identify the most useful tool for vulnerability assessment. Various methods which provide information about groundwater contamination are developed like GOD (Foster, 1987) DRASTIC (Aller et al., 1987), SINTACS (Civita and De Maio, 1997) and so on. These methods have been implemented in different regions over the

Sahajpreet Kaur Garewal, Assistant Professor, NIT Raipur, G.E Road Raipur 492010, India,
e-mail: sahaj012@gmail.com
Avinash D. Vasudeo, Associate Professor, VNIT Nagpur, South Ambazari Road, Nagpur 440010, India

https://doi.org/10.1515/9783110721355-006

world to evaluate the vulnerable areas for effective groundwater planning and management. The methods used for vulnerability assessment are mostly region-specific which vary under different hydrogeological conditions. DRASTIC method is popular tool for assessment of groundwater area under different degree of vulnerability among the alternative available (Babiker et al., 2005; Gupta, 2014; Hamutoko et al., 2016; Rahman, 2008).

Due to increasing urbanization in Nagpur city, overexploitation and quality deterioration of groundwater have been reported in many parts within the city. The city is named after its main river, Nag, which flows within the city and causes adverse impact on the drainage system. Due to disposal of untreated city waste directly into the water bodies, city rivers are getting converted into sewers (Jain and Sharma, 2011). Previous research say that city is affected by higher concentration of contaminant mainly near area adjoining the Nag River and city waste disposal site Bhandewadi (Pujari and Deshpande, 2005; Pujari et al., 2007; Jain and Sharma, 2011).

For sustainable development of groundwater in the city, groundwater vulnerability assessment is essential which can help to identify the areas under high risk of contamination. In the current study, groundwater vulnerable zone of the Nagpur city is assessed using DRASTIC method (hydrogeological parameters) and Multi-Criteria Analysis (MCA; quality parameters).

6.1.1 Study area

The city Nagpur is situated at the geographical centre of India. It lies between 79°00′–79°15′ East longitude and 21°00′–21°15′ North latitudes. The city covers a total area about 218 sq. km within municipal boundary and has a population of approximately 2,405,665. Two major rivers flowing across the city are Pilli and Nag rivers. The geological region of the city is covered by the Archeans and Deccan trap formation (Manzar, 2013). Massive basalt occupies a major portion of the city. Nagpur endure extreme climatic condition, having very hot summer where temperatures rises to 45 °C–48 °C and cold winter having temperature drop to 12 °C–8 °C.

6.2 Material and methodology

Vulnerability of groundwater is estimated considering known condition of aquifers. Each parameter used for vulnerability assessment has its significant impact on groundwater contamination. To define the groundwater vulnerability of the city, the hydrogeological and quality data from government organizations are collected (Table 6.1). ArcGIS 10.0 software is used to prepare all the parameter maps and vulnerability assessment of the Nagpur city.

Table 6.1: Data required for the study.

Data	Organization
Depth to water table, groundwater quality parameters and bore log data to define litho log	Central Ground Water Board, Nagpur (CGWB)
Digital Elevation Model	Bhuvan
The soil map of city in scale of 1:50,000	National Bureau of Soil Survey
Rainfall data	India Meteorological Department.

6.2.1 DRASTIC

The Intrinsic Vulnerability of an area can be determined using DRASTIC method involving known factors of aquifer affecting the transport and attenuation of contaminants. Vulnerability of an area is estimated including the hydrogeological properties of an aquifer (Aller et al., 1987) such as follows:

Depth to water table [D],
Recharge [R],
Aquifer media [A],
Soil Media [S],
Topography [T],
Impact of vadose zone [I],
Hydraulic conductivity [C].

Each parameter is classified to various sub-parameters depending on their characteristics. Rates are assigned to each sub-parameter from 1 to 10 based on the factors they are contributing to contamination. Similarly, on the basis of parameter, importance on vulnerability assessment weights are allocated to each parameter from 1 to 5. Vulnerability Index (VI) is assessed using [eq. (6.1)] by linear summation of rated parameter with individual assigned weight:

$$VI = (D)_r(D)_W + (R)_r(R)_W + (A)_r(A)_W + (S)_r(S)_W + (T)_r(T)_W + (I)_r(I)_W + (C)_r(C)_W \quad (6.1)$$

where, W and r are the weight and rate assigned to the parameters.

The numerical rating and weights documented by Aller et al. (1987) using Delphi technique have been used worldwide and is adopted in the study (Table 6.2) for the assessment of groundwater vulnerable zones. The classification of parameters is tabulated in (Table 6.2) and details regarding impact of each intrinsic parameter and its classification on groundwater is explained in (Aller et al., 1987).

Table 6.2: Rates and weight of the parameters.

1. Depth to the water table (m)			2. Recharge (mm)			3. Aquifer-media		
Sub-parameter	r	w	Sub-parameter	r	w	Sub-parameter	r	w
0.7–2.61	10		396–404	3		Intertrapean	1	
2.62–3.87	9		405–410	5		Basalt (Massive)	4	
3.87–4.69	8		411–416	7	4	Amgaon Gneiss complex	7	3
4.69–5.95	6	5	417–422	8		Unclassified Gneiss tirods	8	
5.95–7.86	4		423–433	9				
7.86–10.77	2							
10.77–15.2	1							

4. Soil media			5. Topography (%)			6. Impact of Vadose zone(m)			7. Hydraulic-Conductivity (m/s)		
Sub-parameter	r	w	Sub-parameter	r	w	Sub-parameter	r	w	Sub-parameter	r	w
			<2.7	10		0.6–3.2	8				
			2.7–5	9		3.2–3.9	7		$<10^{-6}$	5	
Clay loam	3		5–7.9	7		3.9–4.5	6		10^{-5}–10^{-6}	6	
Clayey	7	2	7.9–11	5		4.5–5	5		10^{-4}–10^{-5}	8	
Alluvial	8		11–16	4	1	5–5.9	4	5	10^{-3}–10^{-4}	9	3
			16–23	3		5.9–7	3				
			23>	1		7–10.8	2				

*r = rating and w = weight

6.2.2 Multi-criteria analysis

Three major steps are involved in Multi-Criteria Analysis (MCA): (a) the choice of factor and constrain with following detail of raster maps taking into account their spatial distribution, (b) assigning weight to each factor and (c) the preparation of final MCA map including the weighted factors and limitation.

6.2.2.1 Choice of factors

The factors selected here are variables that show values on a continuing scale. In the current study, groundwater quality parameters are selected factors, and the data of same are collected from CGWB, Nagpur. As we know extensive range of contaminants are present in groundwater and it is difficult to involve each of them. In this study, the groundwater quality data are compared with Indian standard of drinking water and the parameters whose 10% of the samples exceed their permissible limit are selected for analysis.

Quality parameters maps was prepared using (Melloula and Collin, 1998) methodology, by relating the field measured quality parameters value with the desired standard value. The relative values (Vij) for parameters are calculated using [eq. (6.2)]:

$$V_{ij} = \frac{P_{ij}}{P_{id}} \tag{6.2}$$

where, P_{ij} is field quality data of parameter i in cell j and P_{id} is the Indian drinking water standard of parameter i.

The field quality data are transformed using polynomial 2nd order [eq. (6.3)]. The index rating of the quality parameter (Y) will be around 1 for good quality of water, with V_{ij} equal to 0.1. It indicates that $Y_1 = 1$ for $V_1 = 0.1$. Y will be around 5 for acceptable quality of water and 10 for poor water quality which indicates $Y_2 = 5$ for $V_2 = 1$ and $Y_3 = 10$ for $V_3 = 3.5$. It was observed that usually the value of index rating (Y_i) lies from 1–10.

$$Y_i = -0.712V_i^2 + 5.228\ V_i + 0.484 \tag{6.3}$$

where, for any value of V_i, Y_i can be evaluated using parabolic equation.

6.2.2.2 Weight for the parameters

The weight assigned to quality parameters are estimated using Analytic Hierarchy Process (AHP) approach by comparing the parameters on the basis of their impact on human health. AHP is Multi-Criteria Analysis used for solving complex problem (Saaty, 1980) and is used in this study to evaluate the weight of the quality parameters using pairwise comparison approach. The criteria of comparison between the quality parameters are their influence on human health and the surrounding habitat. The parameters are compared on standard 9 levels Saaty's scale, where 9 signifies most essential parameter and 1/9 least essential factor. The overall methodology of solving problem using AHP is very well explained by (Saaty, 1980). Using [eq. (6.4)] consistency index (C.I) of matrix is calculated, to evaluate consistency ratio (CR) of the matrix which is CI/RI (RI is known random matrix). CR of the matrix is checked which should be within a threshold value (10%), and if it is failed then the answers is re-examine.

$$C.I = \frac{\lambda_m - n}{n - 1} \tag{6.4}$$

where, λ_m is the Eigen value (largest) of matrix.

6.2.2.3 Groundwater vulnerability assessment using multi-criteria analysis

After the transformation of the field data values, the vulnerability map using MCA is evaluated by summation of the rating index value (Y_i) of the parameter i for cell j by respective weight of parameter i using [eq. (6.5)].

$$MVI = \frac{C}{n} [W_{ri} \ Y_{ri}] \tag{6.5}$$

where, MVI = Groundwater Vulnerability Index using MCA, C = Constant, n, i = number of parameters involved in the analysis, $W_{ri} = W_i/W_m$, W_i = Weight assigned to parameter i, W_m = max. Weight (5), $Y_{ri} = Y_i/Y_m$, Y_i = the rate estimated by [eq. (6.2)] using value of V_i, Y_m = the maximum possible rate for any parameter (10).

6.3 Results and discussion

6.3.1 DRASTIC

In the study area, groundwater vulnerability was acquired by DRASTIC approach. The vulnerability map is categorized into five vulnerable zones showing area under different levels of vulnerability from very low to very high. The vulnerability map shows (Figure 6.2), out of total study area 14.20% and 49.30% lies in very high-high vulnerable zone followed by 19.02%, 10.09% and 7.39% lies in moderate, low and very low vulnerable zone respectively. From the detailed examination of vulnerability map of the city it was noticed that Central-North East zone of the city are having high vulnerability index and is at higher risk of contamination, where South zone of the city is observed at lower risk of contamination.

6.3.2 Multi-criteria analysis

The groundwater quality parameters whose 10% of the samples exceed their Indian standard drinking-water permissible limit are selected for analysis (Figure 6.1). The seven quality parameters selected for preparing vulnerability map are compared with each other to assess the parameter weights (Table 6.3). The criteria of comparison between the parameters are their impact on human health and the surrounding environment.

MCA Vulnerability Map is evaluated using [eq. (6.5)] seven rated quality parameters to which weights are assigned as per their impact on health and surrounding using AHP (Table 6.3). From the close observation of the resultant MVI map (Figure 6.2), it was found that in city the east-west zone is stressed by higher-moderate concentration of contaminant, where in the west zone vulnerability index lies from is very low to

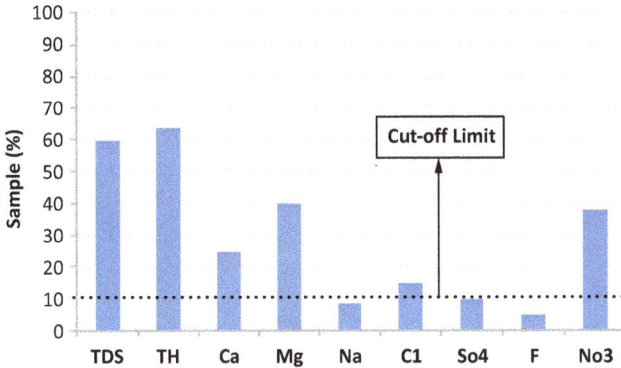

Figure 6.1: Percentage of groundwater samples exceeding permissible limit.

Table 6.3: Pairwise comparison matrix of AHP and weights of the quality parameters.

		1	2	3	4	5	6	7	8	9
		TDS	TH	No_3	So_4	Mg	Ca	Cl	Weights	Re-weight
1	TDS	1	2	1/8	1/6	1/4	1/3	1/2	0.046	0.48
2	TH	1/2	1	1/9	1/7	1/5	1/4	1/3	0.038	0.39
3	No_3	8	9	1	3	5	6	7	0.479	5
4	So_4	6	7	1/3	1	3	4	5	0.203	2.12
5	Mg	4	5	1/5	1/3	1	2	3	0.103	1.07
6	Ca	3	4	1/6	1/4	1/2	1	2	0.075	0.78
7	Cl	2	3	1/7	1/5	1/3	1/2	1	0.057	0.59

moderate and south zones is safe from contamination showing very low vulnerability index value.

6.4 Conclusion

Groundwater vulnerability assessment using traditional approach DRASTIC and MCA is decisive for the incorporation of vulnerability information in the efforts to protect the quality of groundwater regionally. The applied method includes the strength of traditional DRASTIC results and the simplicity and flexibility of MCA. The DRASTIC and MCA occur as analogue approach, where DRASTIC is derived as traditional and the MCA as monitoring method.

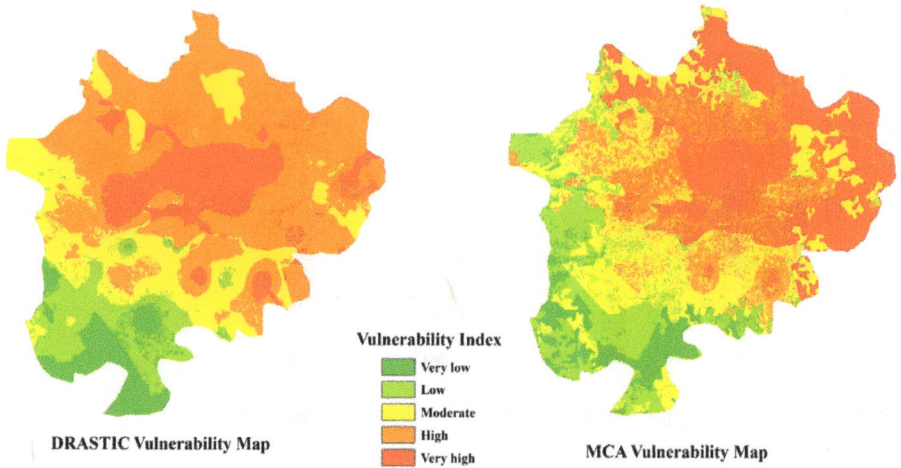

Vulnerability Index

- Very low
- Low
- Moderate
- High
- Very high

DRASTIC Vulnerability Map

MCA Vulnerability Map

Figure 6.2: Groundwater vulnerability map for the study area.

In spite of the slight variation in the results of DRASTIC and MCA, it has been noticed that vulnerability index is least in the south zone of city, and east-north area is highly vulnerable to contamination. This specifies that traditional (DRASTIC) and monitoring (MCA) models are analogous. Therefore, now it can be stated that for a short-time interval and for instant monitoring of aquifer vulnerability MCA can be used. Furthermore, the MCA method, which involves merely groundwater quality parameter for the groundwater vulnerability assessment, provide evidence to be simpler and less time consuming than the traditional model, which involves laborious task of collection and integration of hydrogeological data.

References

Saaty, T. (1980). The Analytic Hierarchy Process, McGraw-Hill, New york.

Pujari, P.R., Pardhi, P., Muduli, P., Harkare, P., Nanoti, M.V. (2007). Assessment of pollution near landfill site in Nagpur, India by resistivity imaging and GPR. Environmental Monitoring and Assessment, 131, 489–500.

Jain, S.K., Sharma, V. (2011). Contamination of Ground Water by Sewage, Central Ground Water Board Ministry of Water Resources Government of India, Faridabad.

Hamutoko, J.T., Wanke, H., Voigt, H.J. (2016). Estimation of groundwater vulnerability to pollution based on DRASTIC in the Niipele sub-Basin of the Cuvelai Etosha Basin, Namibia. Physics and Chemistry of the Earth, Parts A/B/C, 93, 46–54.

Foster, S. (1987). Fundamental concept in aquifer vulnerability pollution risk and protection strategy. Nether-lands: Proc. Intl. Conf.

Rahman, A. (2008). A GIS based DRASTIC model for assessing groundwater vulnerability in shallow aquifer in Aligarh, India. Applied Geography, 28(1), 32–53.

Babiker, I.S., Mohamed, M.A., Hiyama, T., Kato, K.A. (2005). GIS-based DRASTIC model for assessing aquifer vulnerability in Kakamigahara Heights, Gifu Prefecture, central Japan. Science of the Total Environment, 345(1–3), 127–140.

Manzar, A. (2013). Ground Water Information Nagpur District, CGWB, nagpur.

Gupta, N. (2014). Groundwater vulnerability assessment using DRASTIC method in Jabalpur district of Madhya Pradesh. International Journal of Recent Technology and Engineering, 3(3), 36–43.

Johnson, C.C. (1979). Land application of water-an accident waiting to happen. Ground Water. Earth Science. 69–72.

Melloula, A., Collin, M. (1998). A proposed index for aquifer water-quality assessment: The case of Israel's Sharon region. Journal of Environmental Management, 54(2), 131–142.

Pujari, P.R., Deshpande, V. (2005). Source apportionment of groundwater pollution around Landfills site in Nagpur, India. Environmental Monitoring and Assessment, 111, 43–54.

Aller, L., Lehr, J., Petty, R., Bennett, T. (1987). A standardized system to evaluate groundwater pollution using Hydrogeologic setting. The Journal of the Geological Society, 29(1), 23–37.

Civita, M., De Maio, M. (1997). Un sistema parametrico per la valutazione e la cartografia della vulnerabilita' degli acquiferi all'inquinamento, Pitagora Editrice, Bologna.

Ankit Shinde, Vaibhav P. Deshpande

7 Impact of sodium absorption ratio on Kharun river basin, Raipur, Chhattisgarh

Abstract: Sodium adsorption ratio (SAR) plays a vital role in the field of irrigation; its assessment is required for proper management of irrigation. This study involves SAR prediction in the Kharun River in Raipur, southeast of Chhattisgarh, India, using the experimental method that was carried out. The method adopted was to analyse the variation in the SAR value along the flow of the Kharun River. The parameters such as Ca^{++}, Mg^{++}, Na^+, K^+ are the essential parameters in the evaluation of the SAR value of the water sample. The experimental data calculated revealed that there is an increase in the SAR value along the river flow. It was found that the variation in concentration of Na^{++} ion was 3–16.4 mg/L, Ca^{++} ion was 20–130 mg/L and Mg^{++} ion was 50–120 mg/L. The SAR value at the initial was 0.507% to the final value as 1.805% and this variation shows the gradual increase in the SAR value. The SAR value of the river water sample was under the acceptable limit, but there is a gradual increase in its value along the flow. Hence, it can be concluded that there would be an adverse effect to the productivity and yield of the crop if the water at the tail of the river is further supplied to the field for the irrigation purpose. And an effective measure should be taken to reduce the percentage concentration of such ions in the river water during its flow.

7.1 Introduction

Water quality is a significant factor in crop growth, which also contributes to the fertility of the crops. Physical and mechanical properties of the soil, for instance, the composition of the soil (aggregate stability) and permeability are very sensitive to the form of exchangeable ions found in irrigation waters. The chemical analysis can better assess the quality of irrigation water.

The sodium adsorption ratio (SAR) is a parameter used for sodium-affected soil management to maintain the necessary efficacy. Sodium adsorption ratio plays a vital role in growth of plants, and its evaluation is very important so as to use proper water for irrigation. concentrations of the key alkaline and earth alkaline cations in water are checked, which is a deciding factor for the usage of water for the irrigation purpose.

Ankit Shinde, Vaibhav P. Deshpande, Civil Engineering Department, Bhilai Institute of Technology, Raipur, India, e-mail: ankitshinde07@bitraipur.ac.in

https://doi.org/10.1515/9783110721355-007

In water, high concentration of sodium ions influences the permeability of the soil and causes problems related to infiltration. If sodium is present in an exchangeable form in the soil, it removes the calcium and magnesium content of the soil, allowing soil particles to scatter. If the predominant cations absorbed by the soil exchange complex are calcium and magnesium, the soil is good for cultivation having proper permeable and granular structure.

7.1.1 History of Kharun River

Kharun is one of Chhattisgarh's most ancient and historical rivers. Almost 164 km long, the Kharun River flows on its river bank near the capital city of Raipur through the districts of Durg, Bemetara, Dhamtari and Raipur until its confluence in the Shivnath River at the popular pilgrimage spot Somnath of the village of Lakhna near Dharsiwa. One of Chhattisgarh's most ancient and historical rivers is Kharun.

The only source for the supply of water into the canal and streams for irrigation purposes is the Kharun River. The rate of dissolution of waste has risen as a result of the rise in the disposal of solid and liquid waste into the river. This in turn increased the pollution of the water supplied for irrigation purposes to the canals and streams.

7.1.2 Sar value

This is a parameter of irrigation water quality used in sodium-affected soil management, which is calculated from the concentrations of the major alkaline and earth alkaline cations present in the water (Clark et al., 2004; Mise et al., 2007;Pathak et al., 2015; Shah et al., 2013; Asadollahfardi et al., 2013; Cannon et al., 2006). It decides the suitability of water for use in agricultural irrigation. As calculated from the study of pore water present in the soil, it is also a typical diagnostic parameter for a soil's sodium hazard. The SAR measurement formula is as follows:

$$SAR = \frac{Na^+}{\sqrt{\dfrac{Ca^{++} + Mg^{++}}{2}}}$$

In which concentrations of calcium, sodium and magnesium are expressed in milliequivalent/liter. It enables the condition of flocculation or dispersion of clay aggregates in a soil to be evaluated. The dispersion of clay particles is encouraged by sodium and potassium ions, while calcium and magnesium encourage flocculation. In general, higher the value of sodium adsorption level, the less suitable is water for irrigation, while SAR is only one factor in evaluating the suitability of water for irrigation. Irrigating soils using water with a high SAR can permanently change soil structure (Talabi et al., 2017; Tiwari et al. 2017).

7.1.3 Hardness

Water that contains an appreciable amount of dissolved salts of calcium and magnesium. Such salts interfere with the action of soap. The hardness of water sample depends mainly on the concentration of calcium and magnesium ions. Therefore hardness is defined as the concentration of Ca^{2+} and Mg^{2+} ions.

7.1.4 Water sampling

The sampling was done according to the USGS: Standard methods for Water-Quality Sampling. Three consecutive samples were taken from a single location for the relative accuracy in the test results. The figure below depicts the points from were the samples were taken for the further testing. The below picture is a google-generated map showing the locations of the sampling points (Table 7.1 & Figure 7.1)

Table 7.1: Latitude and longitude of sampling points.

Sample no.	Latitude	Longitude
1	Sample of Gangrel Dam	
2	21.3963461	81.6291868
3	21.389176	81.626545
4	21.3613126	81.6078890
5	21.34138	81.5952236
6	21.3333953	81.5758172
7	21.3091062	81.5501598
8	21.2522335	81.5413974
9	21.2459380	81.5449144
10	21.2142720	81.5912898
11	21.2287945	81.5699311
12	21.2083320	81.5977603
13	21.1914599	81.6056665
14	21.1609182	81.6252600
15	21.1488321	81.6073479

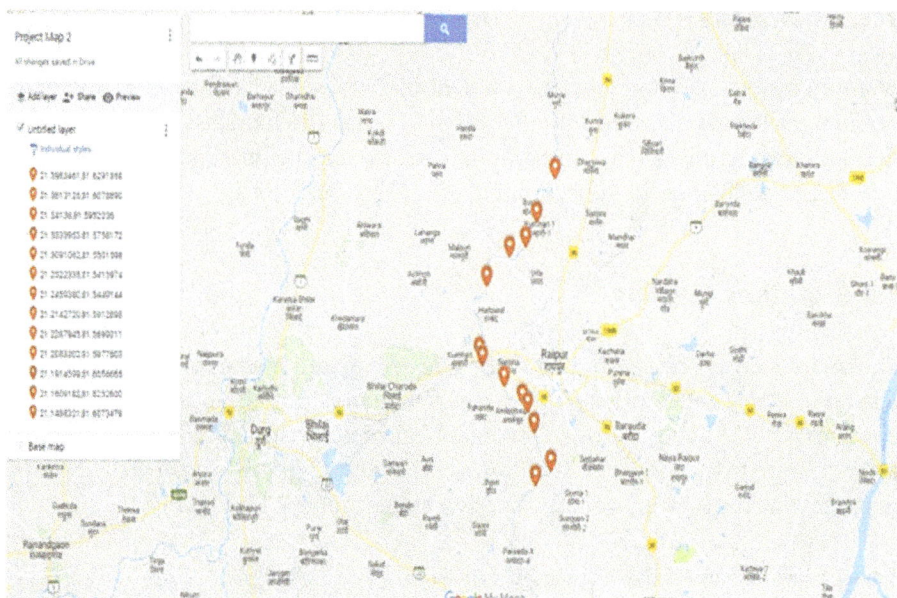

Figure 7.1: Geographic information of sampling points.

7.2 Conclusion

The observations and test results shows an appreciable variation in the data computed. The concentration of various physical and chemical parameters were within the standard limits. But the variation in the value of various parameters were in an increasing order. The concentration of calcium, magnesium, sodium and potassium ions varies from 20 to 130 mg/L, 50 to 120 mg/L, 3 to 16.4 mg/L, 0.1 to 1.4 mg/L, respectively (Table 7.2 & 7.3).

The variation in the SAR value was from 0.507% to 1.805% (Table 7.4 & Figure 7.2) which was under the permissible limit, but since the project area was limited hence there would be a chance of exceedance of the SAR value in the upper basin region.

After Analysing all the observations and data computed, it can be concluded that the SAR value was under the permissible limits but was increasing in nature and can be preferred for irrigation water.

Table 7.2: Hardness.

Sample no.	Calcium (mg/L)	Magnesium (mg/L)
Standard	<200	<100
1	20	50
2	65	45
3	40	85
4	30	35
5	45	25
6	85	55
7	50	50
8	80	35
9	70	55
10	75	60
11	95	75
12	140	45
13	140	20
14	155	120
15	130	35

7.3 Future scope

Since the study area was limited to the Raipur region, the analysis can be done in other river regime of other location and can be analysed in broadened form. Since the variation in the irrigation water quality parameter was increasing in nature, an effective measure should be taken against the prevention of contamination of water regime flowing in the Kharun River (Shukla et al., 2015;Reddy et al., 2013). An analysis can be done on the various effective measures that could be adopted for the betterment of the quality of the Kharun River water.

Table 7.3: Concentration of sodium and potassium ions.

Sample no.	Sodium (mg/L)	Potassium (mg/L)
Standard	<200	<100
1	20	50
2	65	45
3	40	85
4	30	35
5	45	25
6	85	55
7	50	50
8	80	35
9	70	55
10	75	60
11	95	75
12	140	45
13	140	20
14	155	120
15	130	35

SAR value

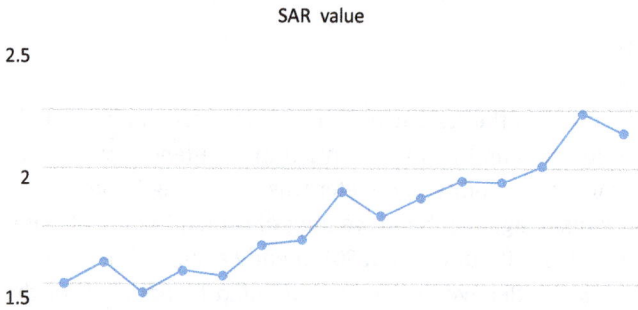

Figure 7.2: Line chart of SAR value.

Table 7.4: SAR value.

Sample no.	SAR value (%)
Standard	<10
1	0.507
2	0.687
3	0.430
4	0.614
5	0.575
6	0.837
7	0.876
8	1.292
9	1.075
10	1.242
11	1.388
12	1.372
13	1.509
14	1.969
15	1.805

References

Civil Engineering Department, Kharazmi University, Tehran Iran. Asadollahfardi, G., Hemati, A., Moradinejad, S., Asadollahfardi, R. Sodium Adsorption Ratio (SAR) Prediction of the Chalghazi River Using Artificial Neural Network (ANN) Iran. (Received: April 25, 2013; Accepted: June 12, 2013)

Cannon, M.R., Nimick, D.A., Cleasby, T.E., Kinsey, S.M., Lambing, J.H. Measured and Estimated Sodium-Adsorption Ratios for Tongue River and its Tributaries, Montana and Wyoming, 2004–06.

Clark, M.L., Mason, J.P. Water-Quality Characteristics, Including Sodium-Adsorption Ratios, for four Sites in the Powder River Drainage Basin, Wyoming and Montana, Water Years 2001–2004

Manoj Kumar Tiwari Associate Professor, Department of Civil Engineering, SSTC – SSGI – FET, Bhilai, C.G India An Analutical Study of Impact of Industrial Effluent on the Kharun River, Raipur, Chhattisgarh.

Mise, D.S.R., Mujawar, S. (Sep -2017). Evaluation of water quality of Kharun River stretch near the Raipur city. International Research Journal of Engineering and Technology (IRJET), 04(09), e-ISSN: 2395-0056, p-ISSN: 2395-0072. www.irjet.net.

Pathak, S.K., Prasad, S., Pathak, T. (Sep, 2015). Govt. College Sanwer, Indore (M.P.) Determination of Water Quality Index River Bhagirathi in Uttarakhand, Uttarakhand India. Social Issues and Environmental Problems, 3(Iss.9: SE). ISSN- 2350-0530(O).

Shah, S.M., Mistry, N.J. (2013). Groundwater Quality Assessment for Irrigation Use in Vadodara District, Gujarat, IndiaWorld Academy of Science. Engineering and Technology International Journal of Biological, Biomolecular, Agricultural, Food and Biotechnological Engineering, 7(7).

Shukla, S., Pandey, D.K., Mishra, D.K. (December 2015). Department of civil engineering, Manoharbhai Patel Institute of Engineering & Technology, Gondia, Maharashtra. Water Quality Assesment of Physiochemical Properties of Shivnath River in Durg District (Chhattisgarh). International Journal of Research in Advent Technology, 3(12). E-ISSN: 2321-9637.

Srinivasa Reddy, K. Assessment of groundwater quality for irrigation of Bhaskar Rao Kunta watershed, Nalgonda District, India Central Research Institute for Dryland Agriculture (CRIDA), Santoshnagar, Hyderabad-500 059, India. Accepted 24 May, 2013.

Talabi, A.O., Afolagboye, L.O., Aturamu, A.O., Olofinlade, S.W. (April 2017). Suitability evaluation of River Owan water for irrigation. IOSR Journal of Environmental Science, Toxicology and Food Technology (IOSR-JESTFT), 11(4), 74–80. Ver. I e-ISSN: 2319-2402,p- ISSN: 2319-2399. www. iosrjournals.org.

Priyanka Tiwari Govt. Arts, Science & Commerce College Tamnar, Raigarh, Chhattisgarh, India, Water quality assesment for drinking and irrigation purpose. Indian Journal of Scientific Research, 13(2), 140–142,2017.

Mahendra Umare, Ashay D. Shende, Valsson Varghese,
A. M. Badar

8 Experimental approach to study the fall velocity of different sized particle in quiescent liquid column

Abstract: The settling velocity of sedimentary particles is one of the most important parameters. Newton and Stokes' law can analyse the settling of particles. If a single sphere is allowed to fall in a liquid media by assuming certain standard conditions, its velocity increases under the action of gravity; but in matter of seconds, the sphere attains a constant velocity under definite boundary conditions. This constant velocity is termed as "Terminal velocity". The particle, under consideration, flowing through liquid is acted upon by following forces as drag force, buoyant force and gravity force. The drag force depends upon the Reynolds number. Fall velocity depends on particle diameter, specific weight of particle, viscosity of liquid media, boundary conditions, temperature and so on. Fall velocity plays an important role in many fields such as design of settling tanks, Water supply project, River Engineering, Irrigation Engineering and so on. In this paper, an attempt through experimentation is made to study the fall velocity of different sized particles in quiescent liquid (water) column. Settling velocity of a particle obtained by introducing a particle in a graduated cylinder containing liquid media. Time required by a particle to settle down is noted. Settling velocity of particle calculated as the ratio of liquid column depth and the time taken by the particle to settle. Experimental value of settling velocity compared with predicted values using an iterative process and plotted the variation between coefficient of drag with Reynolds number.

8.1 Introduction

Newton's and Navier–Stokes equation can analyse the settling of particles. Navier–Stokes equation can be solved for laminar flow around a sphere provided on neglects inertial forces. The solution gives settling velocity of the fall of a sphere in liquid media. Applications of settling or fall velocity are many in Civil Engineering practices (Garde & Raju). A few examples are stated such as Water Supply Project: For the design of settling tank, grit chamber. Fall velocity is an important parameter. River Engineering: For scouring or

Mahendra Umare, Ashay D. Shende, Valsson Varghese, A. M. Badar, Department of Civil
Engineering, KDK College of Engineering, Nagpur-440009, Maharashtra, India,
e-mail: mnu72@rediffmail.com

https://doi.org/10.1515/9783110721355-008

deposition of sediment particle, the settling velocity of the particle in moving water must be known. Irrigation Engineering: Settling velocity is also calculated to know the continuous deposition of sediment in dam, weir so as to decide the life of reservoir.

8.1.1 Background

a) Force acting on particle

When particle starts settling under the action of gravity drag, buoyant and gravitational forces are exerted on the particles.

b) Drag force

The force acting opposite in direction to the velocity of particles in motion is called drag force. Expression for the drag force on spherical objects with very small Reynolds numbers given by Stokes' law.

$$FD = 3\pi\mu dV_f$$

c) Buoyance force

The weight of liquid displaced by the immersed particle is called as buoyant force. It is given as follows:

F_B = weight of fluid displaced = Volume of fluid displaced × density of fluid
= $(\pi d^3 \rho_f g) / 6$ where, d = diameter of a particle & ρ_f = density of fluid

d) Gravitational force

The gravitational force is equal to the weight of particle and it is given as follows:
W = weight of sphere = Volume × density of sphere × g = $(\pi d^3 \rho_s g) / 6$

e) Stokes' law

This calculation was first proposed by Stokes who obtained a solution for general equation of motion by neglecting the inertial terms completely, that is, by assuming that this resistances skin friction. The result of this analysis is generally referred to as Stokes' law and it states that "The force of viscosity on a spherical body falling with sufficiently small constant velocity in a medium is directly proportion to the

Radius of sphere, Fall velocity attained by the sphere and Coefficient of viscosity of the liquid." However, the law is valid under certain assumptions and limited to certain conditions.

$$F = 3\pi\mu dv$$

where, F is the drag force; μ is the fluid viscosity; r is radius of sphere; v is relative velocity between fluid and particle.

f) Factors affecting fall velocity

Effect of internal forces on fall velocity:

K is Stokes' number which can be conceived to be a function of other forces acting on the. Particles can be different from unity (Wasp et al). For R_e more than 0.1, Stokes' law gives result different from those observed in the experiment, because of the influence of inertial forces. Therefore, it can be written in generalized form as follows (Reynolds & Jones 1989):

$$C_D = Re/24 = K$$

When Re is more than 0.1 than inertial forces start becoming significant, one would expect K to be function of Reynolds number. Ossen took into account some of the inertial terms of Navier–Stokes equation and proposed the following equation:

$$C_D = 24/R_e(1 + 3/16R_e)$$

Therefore, $K = R_e(1 + 3/16R_e)$, where R_e = up to

g) Effect of particle size on fall velocity

Stokes' law is applicable to sphere between about 0.2 mm and 0.002 mm in diameters, falling freely through water and having specific gravity close to that of soil grains. However, particle above the range of Stokes' law violate the assumption of the Stokes' law. Rubey in 1953 has suggested a formula for the fall velocity of coarser material, which falls beyond, stokes range. He stated that the total resistance to the motion of particles is the sum of viscous resistance, that is drag force and the impact resistance that is the buoyant force.

Hence, $\pi d^3(\gamma s - \gamma f)/6 = 3\pi d\mu V + \pi d^2\rho_f V^2/4$

Therefore, the fall velocity,

$$V = \text{Sqrt}[(36\mu^2/\rho f^2\ d^2) + (2/3(\gamma s - \gamma f)d/\rho_f)] - 6\mu/\rho_f d$$

h) Effect of particle shape on fall velocity

Stokes assumed that the particle is spherical in shape, but the particles are never perfect spheres. There are either flake-like or needle-like.The effect of particles' shape on the fall velocity can be conveniently discussed by considering two ranges of Reynolds number, namely, Reynolds number less than 0.1 and Reynolds number greater than 0.1. It has been found that for sphere and discs the relationships between CD and Re in stokes range are as follows:

$$\text{For Spheres} \quad CD = 24/R_e$$

$$\text{For Discs} \quad CD = 20.37/R_e$$

These relationships are equally same. From this, we can conclude that drag coefficient is independent of thickness of the particle provided the particle is not relatively not long. Albertson & Schulz, Wilde & Albertson have found that the shape of the particles influences the fall velocity when the Reynolds number is large. They have shown that $c/\sqrt{ab}$ is the shape factor, which influences the fall velocity of coarse particle with Re 10 to 20,000. It has been found that neglecting the shape factor of a particles can give fall velocity which is as much as 300% in error from its true value. According to Schulz, Wilde and Albertson, the range of shape factor is from 0.4 to 0.8 with an average of 0.6. In this theory $c/\sqrt{a^*b}$ is considered as the third parameter while plotting graph between C_D versus Re.

Studies made by Alger and Simons stated that this shape factor $c/\sqrt{a^*b}$ is insufficient in accounting the effect. So they introduced another parameter to this factor as $(c/\sqrt{a^*b})$ dA/dn (Alger & Simons), where dA is the diameter of sphere having same surface area as that of the particle and dn is the nominal diameter of the particle under consideration. Various shape factors are shown in figure 8.1(a) and 8.1(b).

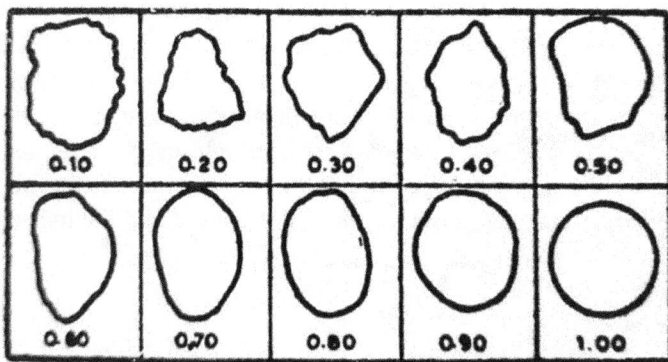

Figure 8.1(a): Shape factor for different irregular particles (Garde & Raju).

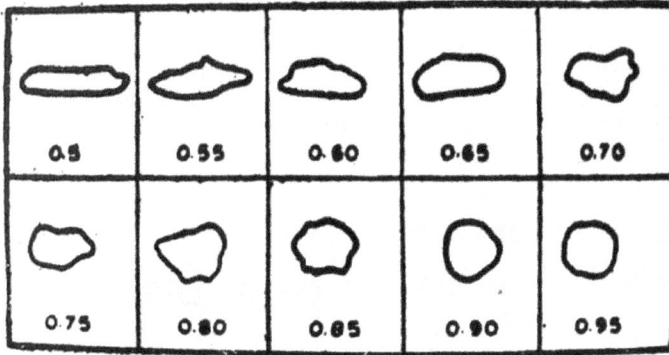

Figure 8.1(b) (continued)

Haider and levenspiel used the term sphericity for various isometric particles to consider their particle shape properties (Gabitto & Tsouris)

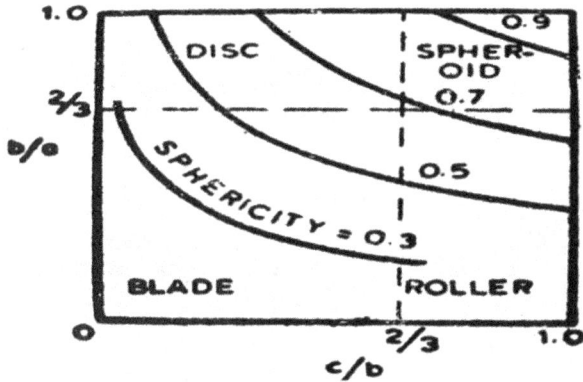

Figure 8.2: Sphericity for different regular shaped particle (Garde & Raju).

i) Effect of wall on fall velocity

The assumption of an infinite extent of fluid in the derivation of Stokes' law is not satisfied in practice and therefore considerable research has been devoted to study the effect of wall on fall velocity.

j) Effect of turbulence on fall velocity

Stokes assumes that the particle fall in calm liquid, that is laminar flow. However, if the Reynolds number is large, flow will be turbulent and the turbulent velocity fluctuations will no doubt have some influence on the fall velocity. Unfortunately, very little work has been done on this subject.

8.2 Prediction of fall velocity

Prediction of fall velocity of the particles has been attempted using analytical method in which an iterative process suggested by Richardson and Zaki has been implemented and the values are compared with the experimental results. Coefficient of Drag and Reynolds Number is considered when the error between assumed preceding Re and existing Re value falls below 5%.

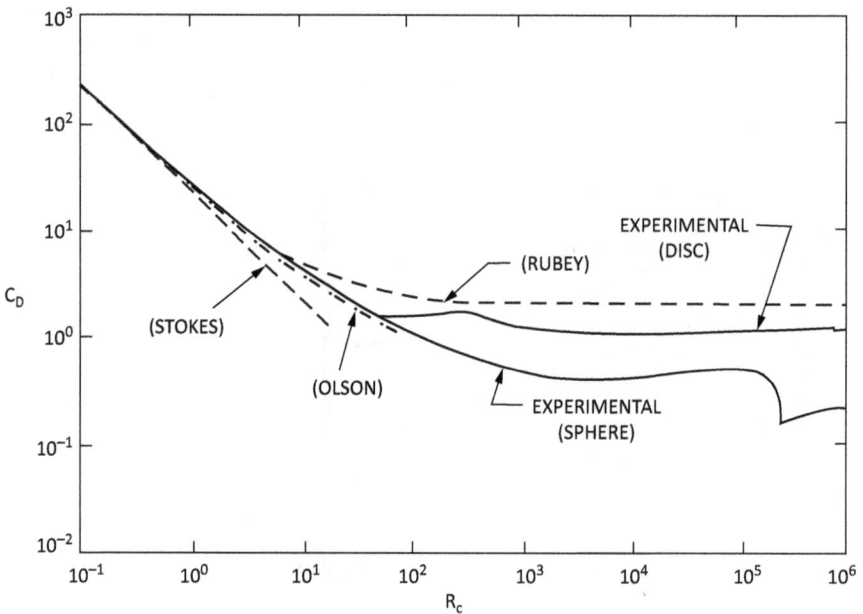

Figure 8.3: C_D versus R_e standard cuve by various researchers.

8.2.1 Experimental setup: water column method

Equipment required: A liquid column, Scale, Stopwatch, 212-micron sieve.
Material needed: Sand, Iron Particles, Liquid Media (Water).
A liquid column is placed on a stable platform/bench. Two points say A & B will be marked on the liquid column.
Length of the water column = 95 cm; diameter of water Column = 5.5 cm and volume of water column = 2.257 L.

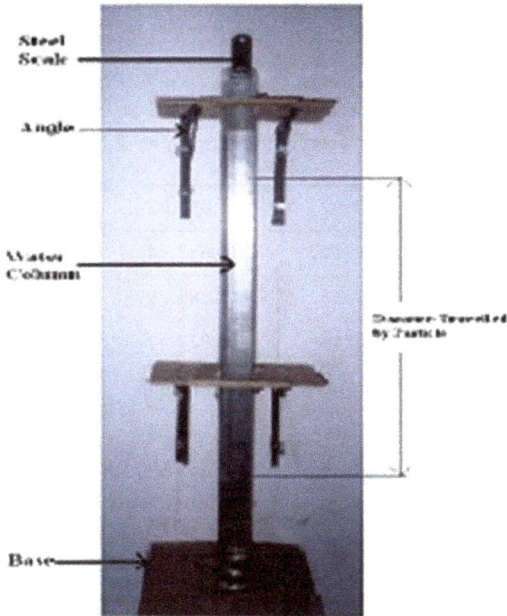

Figure 8.4: Experimental setup.

8.2.2 Richardson and Zaki method

Richardson and Zaki method for prediction of fall velocity of the particle is based on trial and error techniques till the difference between the final value of R_e assumed R_e is less than 5%.

Input Data

Material	Dia1 (cm)	Dia2 (cm)	Dia3 (cm)	Dia4 (cm)	Dia5 (cm)
GRAVEL	0.8682	0.7182	0.7094	0.7090	0.6980
COPPER	0.2744	0.2672	0.2502	0.219	0.1774
BRASS	0.3714	0.3374	0.3274	0.3204	0.2566

8.3 Results and discussion

8.3.1 Experimental approach

Particle	Dia (mm)	Velocity (m/s)	C_D	R_e
GRAVEL	8.682	0.4666	.00592	4,051.02
	7.182	0.3667	.009113	2,633.64
	7.094	0.4063	.00833	2,882.29
	7.090	0.4545	.00745	3,222.41
	6.980	0.3333	.01032	2,323.434
COPPER	2.744	0.4286	.0204	1,176.08
	2.672	0.3667	.02449	979.82
	2.502	0.3571	.026	922.03
	2.190	0.3889	.0282	851.69
BRASS	1.744	0.3250	.04234	566.8
	3.714	1.2500	.0052	4,642.5
	3.374	0.7222	.0099	2,346.7
	3.274	0.7222	.00845	2,682.25
	3.204	0.6500	.01152	2,086.6
	2.566	0.6364	.0151	1,594.82

8.3.2 Richardson and Zaki method

Particle	Dia (mm)	Velocity (m/s)	C_D	R_e
GRAVEL	8.682	0.657	5,455.11	0.44
	7.182	0.597	4,288.52	0.44
	7.094	0.543	4,209.95	0.44
	7.090	0.593	4,206.38	0.44
	6.980	0.589	4,108.87	0.44
COPPER	2.744	0.864	2,206.8	0.44
	2.672	0.794	2,120.6	0.44

(continued)

Particle	Dia (mm)	Velocity (m/s)	C_D	R_e
	2.502	0.78	2,014.43	0.44
	2.190	0.729	1,571.5	0.44
	1.744	0.641	1,072	0.44
BRASS	3.714	0.913	3,390.9	0.44
	3.374	0.87	2,936.09	0.44
	3.274	0.86	2,806.52	0.44
	3.204	0.848	2,717	0.44
	2.566	0.75	1,879.4	0.44

8.3.3 Considering wall effect

Particle	Dia (mm)	Velocity (m/s)	C_D	R_e
GRAVEL	8.682	0.98	0.00472	4,315.67
	7.182	0.95	0.00586	3,474.8
	7.094	0.596	0.00945	2,155.73
	7.090	1.27	0.0095	2,106.84
	6.980	0.95	0.0106	1,958.75
COPPER	2.744	0.742	0.0196	1,039.28
	2.672	0.733	0.0204	998.53
	2.502	0.715	0.022	925.9
	2.190	0.681	0.029	759.64
	1.744	0.619	0.0364	559.62
BRASS	3.714	0.803	0.0134	1,521.15
	3.374	0.77	0.0154	1,322.73
	3.274	0.77	0.0122	1,697.5
	3.204	0.765	0.0163	1,249.7
	2.566	0.707	0.022	925.91

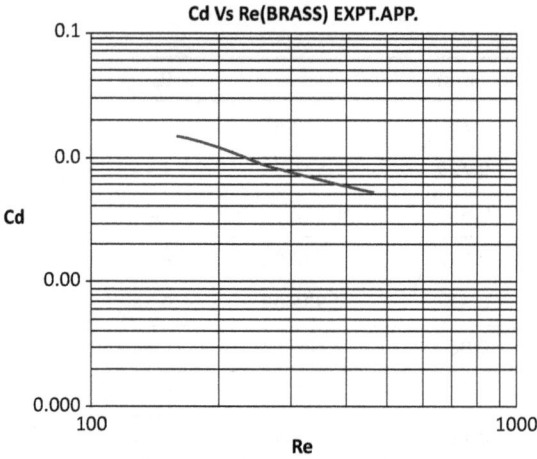

Figure 8.5: C_D Vs R_e graph by experimental method for copper.

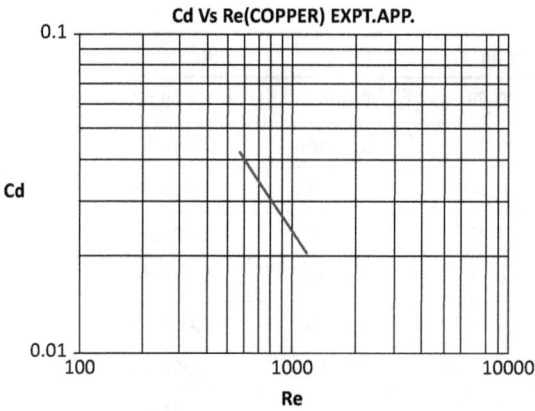

Figure 8.6: C_D vs R_e graph by experimental method for brass.

8.4 Conclusion

Following conclusion are drawn from the above work:
- The value of C_D is found to be 0.44 by Richardson and Zaki method which is higher than the obtained value of C_D by practical approach.
- The actual velocity which is obtained practically is less than the velocity obtained by Richardson and Zaki method. However, the velocity obtained theoretically and considering wall effect is also less than Richardson and Zaki method.

Cd Vs Re(GRAVEL) EXPT.APPR.

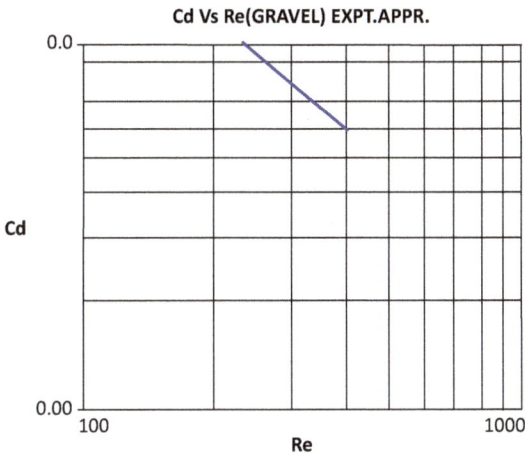

Figure 8.7: C_D vs R_e graph by experimental method for gravel.

GRAVEL

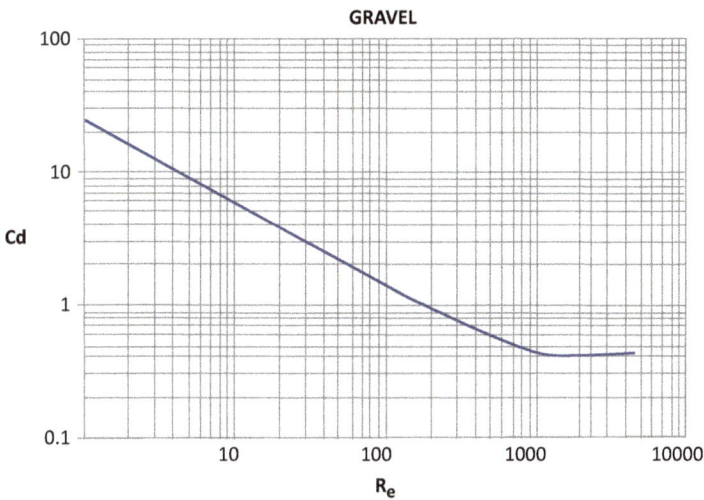

Figure 8.8: C_D vs R_e graph by Richardson and Zaki method for gravel.

This shows that the velocity of particulate is affected by wall boundary wall (since diameter of water column is less (55 mm)).

– The particulate of high mass density and low mass density body is dropped from the same height, the fall velocity (or terminal velocity) of the particulate having high mass density will be too much as the drag per unit weight is high and vice versa.

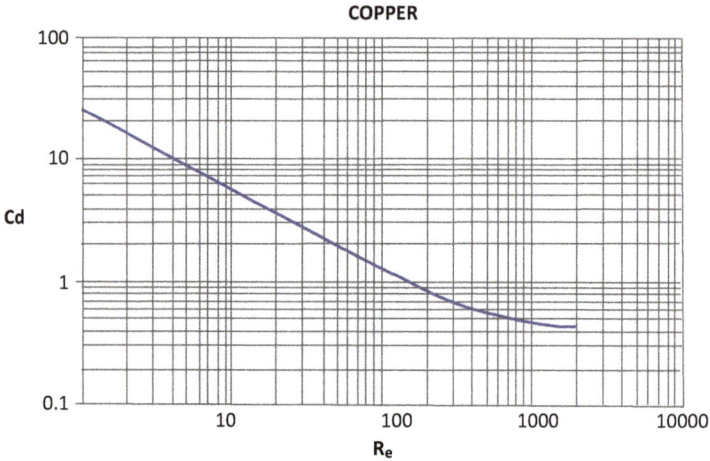

Figure 8.9: C_D vs R_e graph by Richardson and Zaki method for copper.

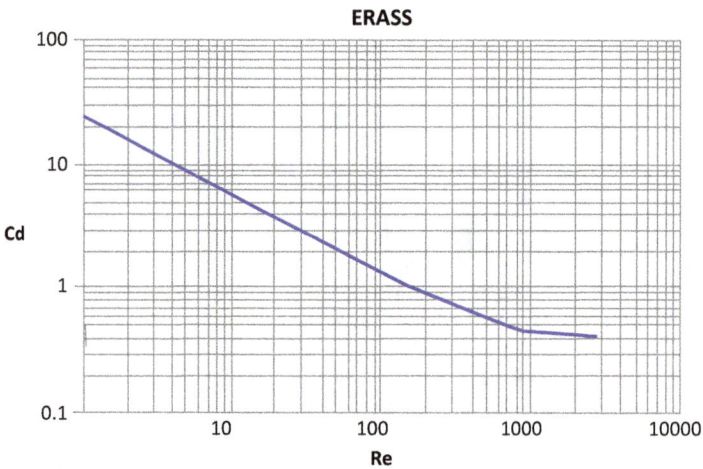

Figure 8.10: C_D vs R_e graph by Richardson and Zaki method for brass.

- Fall velocity is also depend upon the projected area. Since projected area is more, drag force also increases which slowed down the fall velocity by certain amount.
- From combined graph of practical approach it is found that the particulate of same material but of different size have different C_D and R_e values. This shows that the projected area influence the C_D and R_e (e.g., sand and coarse sand, copper and copper alloy).
- From graph by considering wall effect and theoretical approach, we find that the particulate having more C_D values due to wall effect and less C_D in

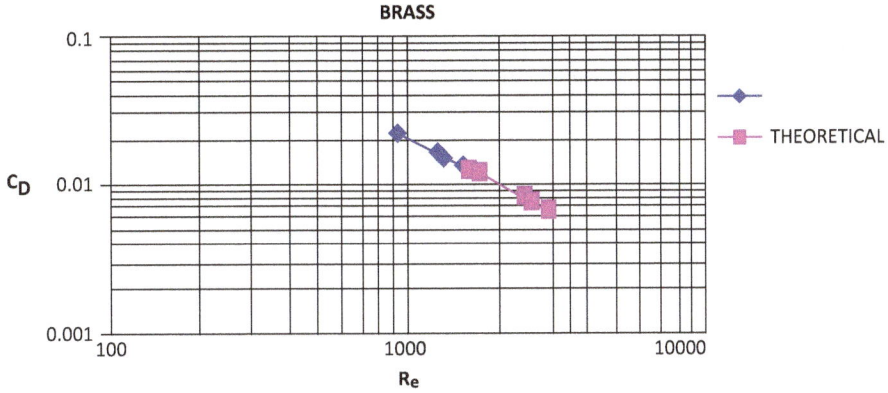

Figure 8.11: C_D vs R_e graph by considering wall effect for brass.

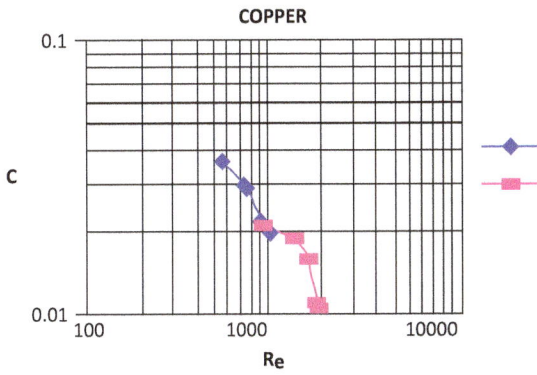

Figure 8.12: C_D vs R_e graph by considering wall effect for copper.

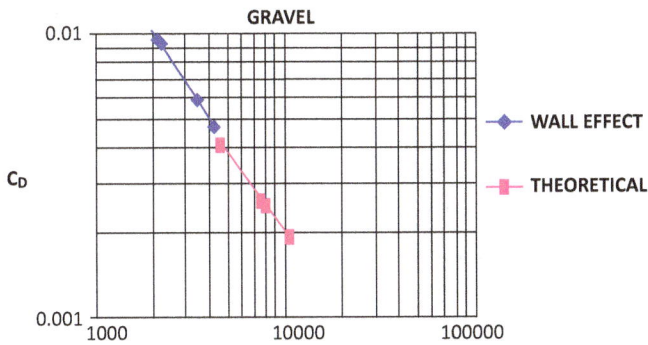

Figure 8.13: C_D vs R_e graph by considering wall effect for gravel.

theoretical approach. Due to wall effect more drag force is acted on particulate whereas in theoretical approach all ideal condition is assumed.

– From graphs plotted by Richardson and Zaki approach, we find that the curve follows same pattern as the standard curve but value will be different because of different types and sizes of material.

– From combined graph by Richardson and Zaki method, we find that when Reynolds number ranges approximately from 1 to 700, the low density particle (e.g., sand, gravel etc.) are above high density particle and then high density particle are above low density particle when R_e is more than 700 for small horizontal distance on graph.

References

Alger, G.R., Simons, D.B. Fall velocity of irregular shaped particles. Journal of Hydraulics Division Proceedings of American Society of Civil Engineers, (M. ASCE).

Effect of Shape on The Fall Velocity of Gravel Particle. Proc. 5th Hydraulic Conference, Iowa 1952.

Effects of particle shape on settling velocity at low reynolds number. Transactions AGV, 31, 1950.

Fall velocity of irregular shaped particles. Journal of Hydraulics Division, Proceedings of American Society of Civil Engineers ASCE, 94(HY-3), May 1968.

Gabitto, J., Tsouris, C. (9 April 2008). Drag coefficient and settling velocity for particles of cylindrical shape. Powder Technology, 183(2), 314–322.

Greenspan, H.P., Ungarish, M. (December 1982). On hindered settling of particles of different sizes. International Journal of Multiphase Flow, Elsevier(6), 587–604.

Hydraulics and Fluid Mechanics By Dr. P.N.Modi and Dr. S.M.Seth, Standard Book House.

Mechanics of Sediment Transportation By Garde and Rangaraju, (Origin and Properties of Sediments)

Mei, R. (April 1994). Effect of turbulence on the particle settling velocity in the nonlinear drag range. International Journal of Multiphase Flow, 20(2), 273–284.

Reynolds, P.A., Jones, T.E.R. (January 1989). An experimental study of the settling velocities of single particles in non-Newtonian fluids. International Journal of Mineral Processing, 25(1–2), 47–77.

Settling velocity of spherical particles in calm water. Journal of Hydraulics Division, Proceedings of American Society of Civil Engineers ASCE, 107(HY-10), October 1981.

Surface area effect on settling velocity of regular geometrical particles. Journal of Institute of Engineers (India), C-1, 62, July 1981.

Tsakalakis, K.G., Stamboltzis, G.A. (March 2001). Prediction of the settling velocity of irregularly shaped particles. Minerals Engineering, 14(3), 349–357.

Wasp, E.J., Kenny, J.P., Gandhi, R.L. Solid- Liquid Flow Slurry Pipeline Transportation, The Relative Motion of Fluids and Particles, First Edison 1977, Trans Tech Publications.

Shubham Chandrakar, Manish Kumar Sinha

9 Comparative analysis of NDVI and LST to identify Urban Heat Island effect using remote sensing and GIS

Abstract: In the age of climate change, we have been significantly witnessing increase in temperature in urban areas, which is one of the major factors influencing urban climatic condition. In cities, the built-up area has low albedo and retains heat for longer period, which increases the temperature and makes the island of high temperature intensity as compared to their surrounding (per-urban area). In urban area, it has been seen that the rainfall pattern has been significantly affected. The role of temperature in urban area is highly important for rainfall occurrence. The rainfall intensity, duration and frequency are no longer similar to the previous patterns in response to urban heat. Thus, the effect of Urban Heat Island (UHI) is being greater challenge to measure, manage and mitigate. This paper addresses the identification of UHI effect by analysing the temperature of land surface that is, Land Surface Temperature (LST), and vegetal area in normalized form that is, normal difference vegetation index (NDVI), durung four different years using temperature bands of Landsat Data on GIS environment. This study has been conducted at capital city of Chhattisgarh state, Raipur City in India. This study covers four years 1988, 2000, 2008 and 2019 to calculate LST and NDVI of the study area in the winter season of India. After that to find out the relation between LST and NDVI, linear regression analysis has been performed. The results indicate a strong relationship between LST and NDVI, and it also shows the increasing pattern of rate of temperature change in urban areas is higher as compared to their surrounding environment, which may be the cause of decrease in vegetation cover in urban areas. The results may be further utilized by the decision-makers in proper planning of the urban sprawl and sustainably manage the urban heat, specifically in Raipur city.

9.1 Introduction

Sustainable development indicates the development without causing harmful effect on environment. Here, this paper presents the analytical study of impact of development (by replacing vegetation cover by built-up area) on the environment also by means of comfort of people regarding the high temperature. In the age of climate

Shubham Chandrakar, Manish Kumar Sinha, Department of Environmental and Water Resources Engineering, UTD, CSVTU, Bhilai, India, e-mail: shubhamchandrakar386@gmail.com

https://doi.org/10.1515/9783110721355-009

change, we have been witnessing increase in temperature in urban areas, which is one of the major factors influencing urban climatic condition. In cities, the built-up area has low albedo and it retains heat for longer period which increases the temperature and makes the island of high temperature intensity as compared to their peri-urban area. Mills (2008) explained the Climate of London is different in temperature of their selected three sites which are outside of London (rural) and one site is in London itself (urban). Earlier this systematic urban climate measuring is called the UHI effects. Boyd (2015) explained the canopy layer of urban and rural areas and the temperature variation on the atmospheric temperature. Canopy layer of atmosphere helps to understand the range of temperature with the altitude, which makes easier to get idea about the intensity of trapped temperature on atmosphere.

Urban Heat Island (UHI) effect is mainly caused by the change in land surfaces which directly affects the Land Surface Temperature (Roth, 2013). Impact of UHI is undesirable as it increases the temperature that directly affects the environment and human comfort (Fernando and Authority, 2017). Due to urbanization, the vegetation cover is replaced by built-up area. Vegetation cover absorbs the solar radiation but utilized immediately; in other hand, built-up area absorbs the solar radiation in the form of heat for longer period which introduces the UHI. As the increase in heat for the cooling of living environment (in the buildings) uses Air Conditioners and refrigerator accelerated, that is it directly affects the use of electricity. Higher use of electricity required greater power grid in the terms of additional power plants that emits carbon dioxide. Also use of AC and refrigerator emits the chloro-flouro-carbon (CFC) gases which enhance the greenhouse gases (GHG), and emission of such gases pollutes the air. Higher temperatures increase the rate of evaporation, combining the process of volatility compounding the long-term human health effect and affect the quality of air.

This study is conducted to analyse the effect of vegetation cover on UHI by analysing the effect of vegetation cover on the Land Surface Temperature (LST). Liu and Zhang (2011) has specified that for the identification of UHI Land surface temperature plays an important role as it shows the variation in temperature on land surface. Bokaie et al. (2016) has explained the trapping of heat at daytime in urban due to multi-storey buildings and streets. Their study also revealed that the extraction of LST is the starting point for the study of UHI. Bokaie et al. (2016) discussed that, for the study of UHI the initial input can be obtained from the thermal imaging of temporal and spatial data of LST. In this study, Landsat TM and ETM + sensor of Landsat satellites is used which provides the LST of the study area. For that the Raipur catchment is used as study area where vegetation cover spreads with lower and higher density so according to their presence in land surface LST has been compared with the Normalized vegetation cover that is NDVI. Initial stage of analysis of UHI is the extraction of LST (Bokaie et al., 2016) as the absorbed heat increases the surface temperature (Coakley, 2003). NDVI shows the health of vegetation as the temperature present in required amount gives green colour, and in excess of temperature it looks brownish. In the comparative analysis of LST and NDVI for identification of UHI, remote sensing is more applicable than any other tools adopted by researchers (Deilami et al., 2018). In this study, Landsat satellite (satellite of United State of Geographical Survey) products were used to extract the LST and NDVI which are freely available at United States of Geological Survey (USGS) earth explorer. Tran et al. (2010) focussed on use of remote sensing for identification of UHI by extracting the LST and normalized indices. Their study has explained the use of thermal band of various satellite data with various resolutions. In this study thermal images were also used to generate LST. Horning (2004) explained the concept of multiband of satellite images and illustrates how individual bands can be used to identify different features on the ground and how these band can be combined to create colour pictures. He also explains about true and false colour. By reviewing previous literature, the data are prepared for this paper.

The entire study depends on remote sensing data which are obtained from USGS's satellite Landsat. In this study, four years' (1988, 2000, 2008 and 2018–19) Landsat image is used for the winter season. For years 1988 and 2008, Landsat 5 satellite having TM sensor; for year 2000, Landsat 7 satellite having ETM + ; and for year 2018–19, Landsat 8 satellite having OLI-TIRS sensor were used. Different Landsat satellite contains different number of bands with different properties. Landsat 5 and Landsat 7 satellites contain 7 and 8 number of bands, respectively and Landsat 8 satellite contains 12 number of bands. Reflective bands NIR, SWIR and visible bands are used to calculate NDVI, NDBI and NDWI, respectively. Also, thermal bands are analysed to extract LST.

9.1.1 Study area

Urban Heat Island effect shown in the urban area is surrounded by nearby periphery urban area or rural area. The surrounding of the Kharun River and vegetation cover of Raipur catchment is sufficient to show the healthy side of vegetation cover and the Raipur city is present to show the adverse effect of UHI. Urban Heat Island effect shown in the urban area which is surrounded nearby periphery urban area or rural area For the identification of UHI, an area which has all features such as built-up area, vegetation cover, water body and bare soil is most suitable. Under Raipur catchment (Figure 9.1), Durg and Raipur district is taken as study area of this study. Both the major cities of Chhattisgarh contribute to the major part of Kharun sub-basin. Coordinate of this study area is 21°5' to 21°25' in north direction and 81°25' to 81°45' in east direction (Kumari et al., 2016) (Sinha et al., 2019). The study area covering an area about 1,118.25 sq km., and the map projection of the study area is Universal Transverse Mercator (UTM) and its zone is Northern hemisphere 44. Projection unit is meters. The datum is World Geodetic System 84 (WGS84). The image of satellite is obtained in tiles from which study area is extracted. For the identification of UHI, an area which has all features like built-up area, vegetation cover, water body and bare soil is most suitable. Under Raipur Catchment (Figure 9.1), Durg and Raipur district is taken as study area of this study. Both the major cities of Chhattisgarh contribute to the major part of Kharun sub-basin. Kharun River is a major tributary of Seonath River which is a tributary of Mahanadi River.

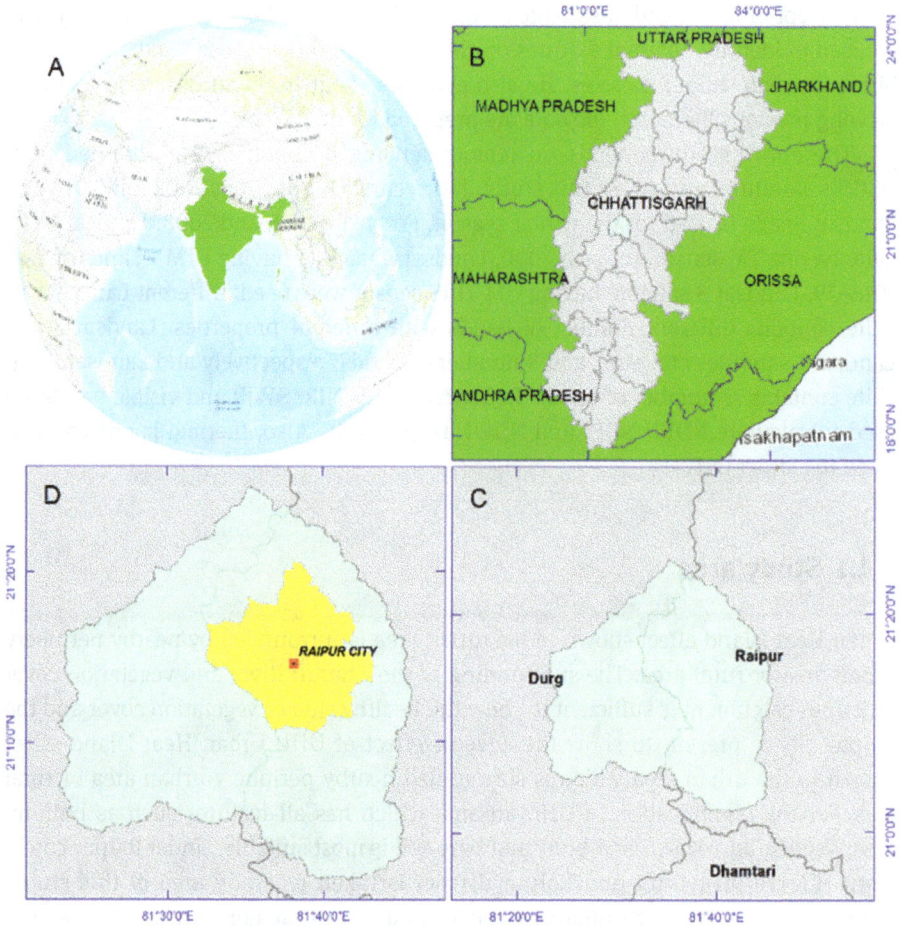

Figure 9.1: Location map of the Raipur catchment.

9.1.2 Methodology

The methodology of this study is done to achieve objectives to identify the effect of UHI. To achieve this objective, calculation of LST and NDVI are done using 3 Landsat satellite (5, 7 and 8) data. This methodology shows how to calculateLST using the Landsat 5 bands. In particular, band 6 is the thermal band, band 3 and 4 are used to calculate the Normal Difference Vegetation Index (NVDI). For this study, 142 number of paths and 45 number of rows are used and the date on which image acquired is on 17 September 1988. It is one of them from which LST had been calculated, the values of all LST from all four years are further listed in the next section. The LST can be estimated or calculated using the Landsat thermal bands with the help of equation which is used later (Pal and Ziaul, 2017) in this methodology. It simply requires applying a set of equations through a raster calculator (Arc Map). The first step is to download a Landsat 8 image from a particular location, unzip it and check certain information needed (within the metadata) to execute this procedure. The process is synthesized in six steps below:

Extraction of Top of Atmospheric (TOA) spectral radiance

$$TOA = \text{Radiance Mult Band} \cdot \text{Thermal band} + \text{Radiance Add Band} \qquad (9.1)$$

For Landsat 5,
    Thermal band = Band 6
    M_L = 5.5375E-02
    A_L = 1.18243

TOA to Brightness Temperature (BT) conversion

$$BT = (\text{Thermal constant K2}/(\ln\,(\text{Thermal constant K1}/TOA) + 1)) - 273.15 \qquad (9.2)$$

For Landsat 5,
    K1 = 607.76
    K2 = 1,260.56
    In this equation 273.15 °C is subtracted to obtain the results in Celsius.

Calculation of NDVI

$$NDVI = \text{float}(\text{Band 4} - \text{Band 3})/\text{float}(\text{Band 4} + \text{Band 3}) \qquad (9.3)$$

This is the modification of equation of Townshend and Justice (1986), which is used in raster calculator (Arc Map) to obtain the normalized value. In the calculation of LST, NDVI is used to estimate the proportion of vegetation (P_v), which is used to calculate the emissivity (ε).

NDVI is used to indicate the vegetation condition on the Earth's surface and also the health condition of vegetation. The chlorophyll content of vegetation absorbs the Red (R) band of sun radiation and reflect the near-infrared (NIR) band of sun radiation. NDVI computed using these two bands of sunlight and the equation formed by Townshend and Justice (1986).

Calculation of the proportion of vegetation

$$P_v = ((NDVI - NDVI_{min})/(NDVI_{max} - NDVI_{min}))^2 \qquad (9.4)$$

In ArcGIS environment at the content of table the minimum and maximum values of the NDVI raster dataset can be displayed directly in the image; otherwise, it is obtained by the properties of that NDVI raster dataset.

Calculation of emissivity (ε)

$$\varepsilon = 0.004*P_v + 0.986 \qquad (9.5)$$

Simply apply the formula in the raster calculator; the value of 0.986 corresponds to a correction value of the equation.

Calculation of land surface temperature

$$LST = (BT/(1 + (\lambda*BT/\rho)*Ln(\varepsilon))) \qquad (9.6)$$

where,
λ = wavelength of emitted radiance, $\rho = h*c/s = 14{,}388$ µm K, h = Planck's constant, s = Boltzman constant, c = velocity of light.

Values of λ are different for different Landsat centre; wavelength of Landsat band is used which is thermal band. For the Landsat 5 and band number 6, the value of λ is 11.45 µm. By applying this equation, land surface temperature map has been obtained which is further classified with their variable range in this study. Obtained surface temperature map is only land surface temperature not the air temperature.

9.2 Results and discussion

Figure 9.2 shows the surface temperature map of LST in four years 1988, 2000, 2008 and 2018 in month of January. In all surface temperature maps, bright reddish colour shows higher temperature and bluish colour shows lower temperature. These surface temperature maps indicate the change of LST pattern which rapidly change. Usually, the range of LST is 19.06–34.27 °C during January 1988. Since 1988–2000, 2000–2008 and 2008–2018 have decreased and increased from 19.06 °C to 18.08 °C, 18.08 °C to 21.73 °C and from 21.73 °C to 18.73 °C, respectively.

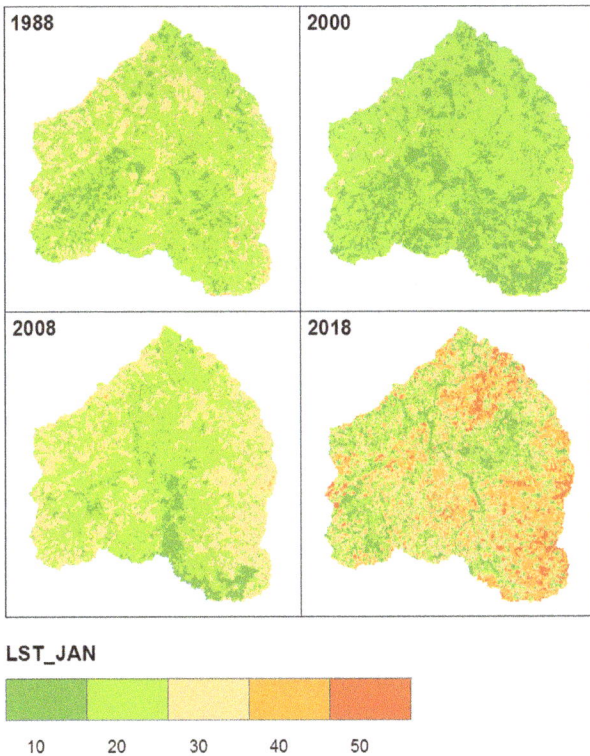

Figure 9.2: Maps of LST for the January month of 1988, 1998, 2008, 2018.

Vegetation cover represent the absorbed solar radiation in the form of heat with immediate utilization, is extracted by NDVI (Figure 9.3). If heat is in permissible amount, it shows the healthy vegetation which reflects greenish colour while excess amount of heat reflects brownish colour which cannot be utilized by leaf of vegetal cover, and so the greenish colour shows the minimum surface temperature and maximum temperature is indicated by brownish colour. To ensure the variation of temperature regarding the land type, that is vegetation land surface temperature has been calculated which also shows the maximum temperature in that land type where the vegetal cover is less (i.e., urban area). Figure 9.3 shows NDVI maps derived from the multi-bands satellite images as explained in Chapter 4. NDVI maps are classified for making a relation between LST and NDVI of different intensity levels at land surface. Figure 9.3 depicts the extracted values of NDVI and classified into five classes which are shown by Seville orange, mango, electron light apple, quetzal green and fir green colour in NDVI map. Seville orange and mango show the rocky surface land and built-up area which absorbs NIR that is clearly seen in NDVI map as it gives less value. Light apple shows the bare soil of study area which gives moderate value of NDVI, while quetzal green and fir green indicate the vegetation including

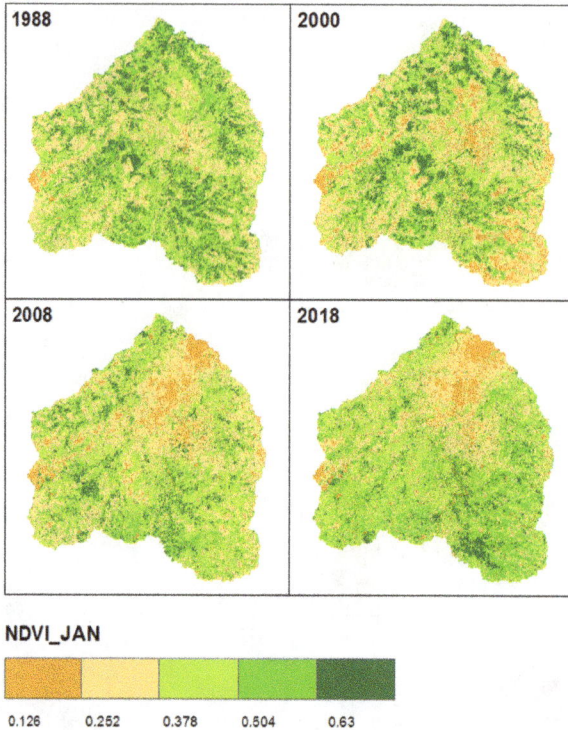

NDVI_JAN

0.126	0.252	0.378	0.604	0.63

Figure 9.3: Maps of NDVI for the January month of 1988, 1998, 2008, 2018.

agriculture land and forest. In all NDVI maps, quetzal green colour shown in rural area and mango colour in urban area which shows the NIR absorption surface leads in urban area. Table 9.1 describes the variation of LST of the selected years, which shows the increase in temperature at overall study area more significantly in urban area which indicate the presence of UHI. Table 9.1 shows the relation between LST and NDVI with numeric value, which shows the dependency of LST on the vegetation cover.

In 1988, the NDVI range is in between 0.237 and 0.685, which gradually decreased between 0.130 and 0.0395 in 2018. Therefore, it can be said that the NDVI decreased in urban heartland over time. As per the documented results work, R^2 value for January 1988 is 0.973, and it is raised to 0.984 in 2018 (Figures 9.4 and 9.5). This result shows decreasing of NDVI value, that means LST increases with the decreasing of NDVI specially in the urban area where the reddish and brownish colour highlighted, which indicate the higher amount of heat in urban area with respect to their surroundings.

Table 9.1: Description of LST variation and relation between LST vs NDVI.

Year	Month	Min. temp in °c	Max. temp in °c	LST vs NDVI
				Coefficient of determination (R^2)
1988	January	19.06	34.27	0.9675
2000	January	18.02	47.85	0.9686
2008	January	21.73	49.77	0.9909
2018	January	22.73	51.68	0.9845

(a)

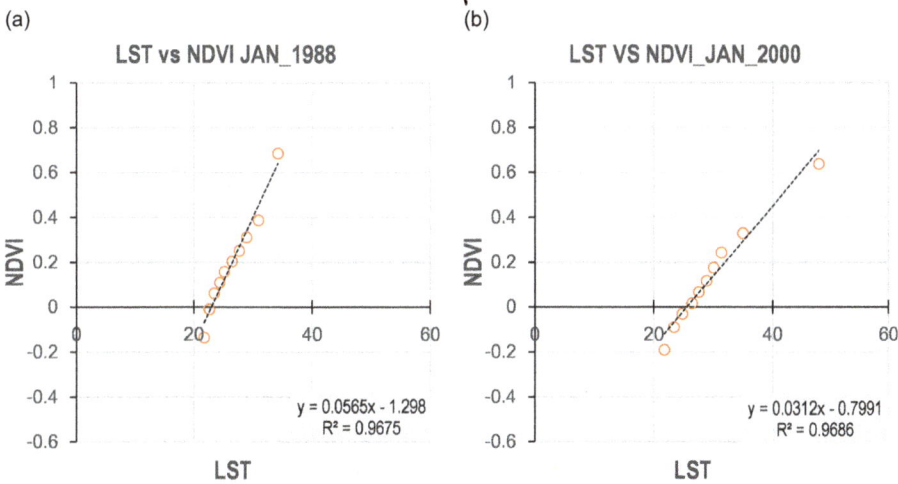

(b)

Figure 9.4: (a) and (b) Relation in LST and NDVI in 1988 and 2000.

9.3 Conclusion

The present scenario of Indian urbanization is assumed to be developed in a smarter way – that is, making of smart cities. The term smart stands to develop the cities in well organized, well planned and more comfortable. The atmospheric suitability (comfort) is affected by the heat, and so for the development of smart city consideration of heat management is also needed for that phenomenon related to heat is required to analyse. The study shows presence of effect of UHI in Raipur catchment. Major loss in vegetated cover in urban area, is required to be controlled during the development of smart cities. Urban Heat Island is basically temperature variation phenomenon which shows the increase in temperature relative to surrounding area. This study covers both temperature variation condition land surface temperature and also atmospheric temperature. By considering the results of study, it is concluded

(a)

LST VS NDVI_JAN_2008

$$y = 0.0344x - 1.0419$$
$$R^2 = 0.9909$$

LST

(b)

LST VS NDVI_JAN_2018

$$y = 0.0365x - 0.7779$$
$$R^2 = 0.9845$$

LST

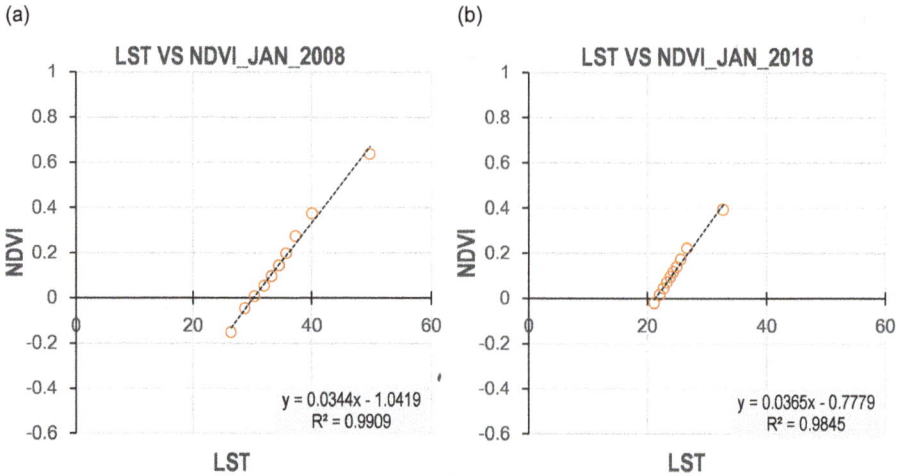

Figure 9.5: (a) and (b) Relation in LST and NDVI in 2008 and 2018.

that the magnitude of increase in temperature is higher in the rural as compare to peri-urban area in Raipur. It is also concluded that the rainfall affected by temperature and the variation in rainfall with the temperature is higher in urban area compared with the rural areas. As per concern of sustainable development, it is necessary to measure, manage and mitigate the adverse impact of development. This paper clearly indicates the adverse effect of replacing vegetation by cause of urbanization which affects the rise in temperature that is UHI. From the results, it can be concluded that the UHI is present at Raipur catchment in the Raipur city which required remedial management and mitigation. By the rise in temperature, UHI also affects the climate and as we know the climate change is already a big trouble for environment and people. This also shows the important of vegetal cover in terms of reducing the temperature in rural area in comparison with urban area.

References

Bokaie, M., Zarkesh, M.K., Arasteh, P.D., Hosseini, A. (2016). Assessment of Urban Heat Island based on the relationship between land surface temperature and land use/ land cover in Tehran. Sustainable Cities and Society, 23, 94–104. https://doi.org/10.1016/j.scs.2016.03.009.

Boyd, K.D. (2015). Identifying enhanced urban heat island convection areas for Indianapolis, Indiana using space-borne thermal remote sensing methods. April.

Coakley, J.A. (2003). Reflectance and Albedo, surface. Encyclopedia of Atmospheric Sciences, 1914–1923. https://doi.org/10.1016/b0-12-227090-8/00069-5.

Deilami, K., Kamruzzaman, M., Liu, Y. (2018). Urban heat island effect: A systematic review of spatio-temporal factors, data, methods, and mitigation measures. International Journal of Applied Earth Observation and Geoinformation. https://doi.org/10.1016/j.jag.2017.12.009.

Fernando, T.S., Authority, C.E. (2017). Identification of relationship between Urban Heat Islands & Vegetation using landsat 8 A case study of Colombo & Gampaha districts in Sri Lanka Department of Geography University of Sri Jayewardenepura. September 2016. https://doi.org/10.13140/RG.2.2.33968.76806.

Horning, N. (2004). Selecting the appropriate band combination for an RGB image using Landsat imagery Version 1.0. American Museum of Natural History, Center for Biodiversity and Conservation, 14. http://biodiversityinformatics.amnh.org.

Kumari, S., Jha, R., Singh, V., Baier, K., Sinha, M.K. (2016). Groundwater vulnerability assessment using SINTACS model and GIS in Raipur and Naya Raipur, Chhattisgarh, India. Indian Journal of Science and Technology, 9(41). https://doi.org/10.17485/ijst/2016/v9i41/99247.

Liu, L., Zhang, Y. (2011). Urban heat island analysis using the landsat TM data and ASTER Data: A case study in Hong Kong. Remote Sensing, 3(7), 1535–1552. https://doi.org/10.3390/rs3071535.

Mills, G. (2008). Luke Howard and the climate of London. Weather, 63(6), 153–157. https://doi.org/10.1002/wea.195.

Pal, S., Ziaul, S. (2017). Detection of land use and land cover change and land surface temperature in English Bazar urban centre. Egyptian Journal of Remote Sensing and Space Science, 20(1), 125–145. https://doi.org/10.1016/j.ejrs.2016.11.003.

Roth, M. (2013). Urban Heat Islands. Handbook of Environmental Fluid Dynamics, 143–159. https://doi.org/doi:10.1201/b13691-13.

Sinha, M.K., Baghel, T., Baier, K., Verma, M.K., Jha, R., Azzam, R. (2019). Impact of urbanization on surface runoff characteristics at catchment scale. Water Resources and Environmental Engineering I, 31–42. Springer Singapore. https://doi.org/10.1007/978-981-13-2044-6_3.

Townshend, J.R.G., Justice, C.O. (1986). Analysis of the dynamics of african vegetation using the normalized difference vegetation index. International Journal of Remote Sensing, 7(11), 1435–1445. https://doi.org/10.1080/01431168608948946.

Tran, A., Goutard, F., Chamaillé, L., Baghdadi, N., Lo Seen, D. (2010). Remote sensing and avian influenza: A review of image processing methods for extracting key variables affecting avian influenza virus survival in water from earth observation satellites. International Journal of Applied Earth Observation and Geoinformation, 12(1), 1–8. https://doi.org/10.1016/j.jag.2009.09.014.

Index

adsorption 28
Analytic Hierarchy Process 71
ArcGIS 68

back trajectory 59

COD 11
Common Effluent Treatment Plant 15

dissolution 31
Districts Resource Map 69

electrical conductivity 43

fall velocity 85
fluorine 18

hardness 79
hydrodynamic cavitation 9
hydrological runoff depth 7
HYMOS 2
HYSPLIT 59

Kharun river 71

latitude and longitude 79
LULC 71

MPN method 45

PM_{10} 59
precipitation 40

Reynolds number 88

settling velocity 85
SWDES 4

total dissolved solids 46

United States Geological Survey 69
Universal Transverse Mercator 101
Urban Heat Island 100

https://doi.org/10.1515/9783110721355-010